T0189910

Czechoslovakia: the Velvet Revolution and Beyond

Czechoslovakia: the Velvet Revolution and Beyond

Robin H. E. Shepherd

palgrave

Published by PALGRAVE
Houndmills, Basingstoke, Hampshire RG21 6XS and
175 Fifth Avenue, New York, N. Y. 10010
Companies and representatives throughout the world

PALGRAVE is the new global academic imprint of
St. Martin's Press LLC Scholarly and Reference Division and
Palgrave Publishers Ltd (formerly Macmillan Press Ltd).

Outside North America
ISBN 978-0-333-79188-2 hardcover
ISBN 978-0-333-92048-0 paperback

In North America
ISBN 978-1-349-62811-7   ISBN 978-1-137-07975-6 (eBook)
DOI 10.1007/978-1-137-07975-6

This book is printed on paper suitable for recycling and
made from fully managed and sustained forest sources.

A catalogue record for this book is available from the British Library.

Library of Congress Cataloging-in-Publication Data
Czechoslovakia : the velvet revolution and beyond / Robin H. E. Shepherd.
p.   cm.
Includes bibliographical references and index.

1. Czech Republic— Politics and government—1993– 2. Czech Republic–
–Economic conditions. 3. Slovakia—Politics and government—1993–
4. Slovakia—Economic conditions. I. Title.
.
DB2244.7 .S44   2000
943.705—dc21   99–051860

10   9   8   7   6   5   4   3   2
08   07   06   05   04   03   02   01

To Pat, Jonathan and the memory of Tony

Contents

List of Tables

Acknowledgements

Many people are owed thanks for encouraging and helping me write this book. David Bolchover read through the first three chapters. Jonathan Stein made several helpful suggestions to Chapter 4. Josef Poeschl was methodical in his perusal of Chapter 5 on the economy, the biggest in the book. Jana Dorotková added several points to Chapter 8 on the potentially hazy subject of post-communist Slovak politics. In a more general sense, Paul Lewis was a source of both encouragement and lively debate. The book has benefited from suggestions by all of the above, but any errors, oversights or weaknesses are the sole responsibility of the author. There are many others who should, but for various reasons cannot, be mentioned. I am grateful to them.

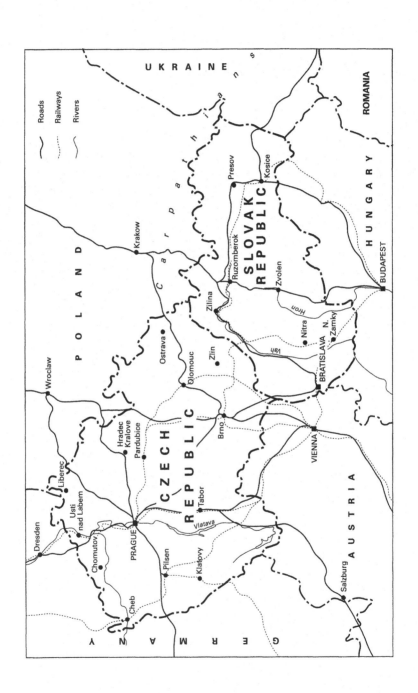

Introduction

The events in Czechoslovakia and elsewhere in eastern Europe in late 1989 were genuinely revolutionary in character. This is so in the sense that the positions of the powerless and the powerful changed such that those who were in jail, threatened with jail or freshly freed from jail took control of the apparatus of state power and those that had held it were reduced to the status of political outcasts. It is also true in the wider sense that the economy and the social base which it engendered were destined for change at the most fundamental level. Finally, it shared in common with other genuine revolutions the deviant characteristic of not in fact being able to entirely remove members of the old elite from the establishment. The former communists who did survive, however, could wield authority as individuals but not as representatives of a power structure which the revolution had demolished.

What followed the revolutions in eastern Europe was bound to be affected by what had gone before. This is, of course, a truism. But it does at least focus our attention on the particularly unusual conditions under which Czechoslovak and other Soviet bloc citizens lived. It simultaneously brings us to the core issue under discussion in this book: How far has the legacy of communism been shaken off and how much remains to be done?

The uniqueness of the system of rule practised by communist regimes was its totalitarian character. The model totalitarian state was developed by Lenin and perfected and completed by Stalin in the Soviet Union in the 1930s. Its main features have generally been subdivided into the following six categories: a single monopoly party with mass support, a socialist economy, a personality cult of the party leader stressing his perfect embodiment of the ideals of the revolution, a ruling ideology giving answers to all social, political, moral and sometimes even scientific questions which posits a utopian end state towards which the world as a whole is moving, a totally integrated and state controlled media to project this ideology, and finally an invasive security police spreading mass terror throughout society.

Neither in the Soviet Union since 1953 nor in eastern Europe at any time since the communist takeovers after the Second World War have these six categories come together with the force they did in Stalin's Russia. The specific difference between Stalinism and the style of rule practised in eastern Europe was the general absence of the leader cult and the relatively much more selective use of terror as an instrument of political control. Nevertheless, the socialist economy, the state media, the invasion of civil society by organs of the communist party and the presence of a ubiquitous informer network gave communist rule a totalising character which contrasts sharply with some other authoritarian regimes in the twentieth century. The consequences of political opposition in Apartheid South Africa, for example, or Pinochet's Chile, could be extremely severe but the aims and the scope of the state were not such that total conformity was either possible or necessarily desirable to the ruling elites.

In both countries, secret and sometimes open political opposition, the provision of alternative and expressly political platforms, did and could occur. Czechoslovakia, one of the most orthodox and efficiently organised of the region's totalitarian regimes, had the ideological will, the state secret police apparatus and the economic and societal dominance to reduce opposition to tiny networks of intellectuals with no political, still less party political, programme save objection to the very system which constrained their lives. The atomisation which accompanied the destruction of civil society also prevented the formation of anything recognisable as a public opinion. People had private views of course. But the absence of any independent organisations, the absence even of organisations existing under severe constraints, prevented the emergence of public opinion in a unified and articulate form.

The significance of all this to the revolutions of 1989 and what came after is broadly two-fold. The revolutions were led against communism but not necessarily in favour of anything in particular (although liberal democracy was the obvious choice as a general framework) and the political groupings which assumed power were disparate coalitions of all those interests, including reformed communists, which desired change. The revolutionaries were long on idealism but short on specifics. Total government engendered total but generalised opposition. In the case of Czechoslovakia (and the two other federations in the communist world) there was an additional problem.

Ten years after communism the most obvious change to the area once called Czechoslovakia is that its constituent nations have split into two separate and independent states. Constructing a capitalist liberal democracy out of the ruins of communism was therefore complicated by the parallel task of reforging the national identities of two relatively small European countries and projecting them to a foreign audience still reeling from the world historical changes associated with the fall of the Soviet Union. The Czech and Slovak peoples were confronted with the problem of a 'triple transition' involving fundamental change to the nation, the economy and the political framework in which both were expressed.

In the early years after 1989 it was probably sensible to discuss the revolution against communism from within this paradigm, explaining how each element of that triangular relationship impacted on the other, eventually leading to the break-up of the state and to some extent colouring immediate developments in both the Czech and Slovak Republics once independence had been achieved.[1] Quite how significant the national question was to the other two aspects of the transformation project, and how important they were to it, is debatable. In retrospect it seems that 'triple transition' was useful as an interpretive device explaining the division of the state itself and subsequent political developments in Slovakia, but less so for the Czech Republic. The Czechs, who tended to view Czechoslovakia as a version of their own nation writ large, appear to have accepted independence with greater ease, viewing it as a diminution in the degree of Czech power but not as a qualitative change to their national and political identity. The validity of this observation is perhaps borne out by the lines of continuity we can see in the Czech political and economic transformation before and after 1993, and the sharp discontinuities evident in Slovakia. Roughly speaking, the Czech Republic after independence followed the pattern established in the months and years after 1989: Václav Havel and his dissident intellectual friends set about the construction of a modern democratic state, while the other Václav, Finance Minister and later Prime Minister, Klaus, sought to push through the most radical free market economic experiment in the former eastern bloc. In these important respects, nothing substantial changed. Nationalism affected minority groups within the Czech Republic but mainstream political and economic reform was not thrown off course because Czechs were now managing their affairs without Slovak help.[2]

The Slovak experience has been very different. At independence, the capital of Czechoslovakia became the capital of the Czech Republic. What was once merely the republican capital Bratislava became the seat of an entirely new source of political power. Slovaks were thrown back on their own resources. With the dubious and damaging exception of Jozef Tiso's Nazi puppet regime in the Second World War, Slovaks had nothing from their own past to guide them in the ways of statehood. Matters were complicated further by the uncertain degree to which the new state could count on the support of its people. Judging precisely what proportion of the Slovak population actually wanted independence was not easy. What was certain was that many, especially among the intellectual elite, were deeply concerned that Slovakia had taken a wrong turn. From the very start, therefore, the country had to contend with the crucial question of who was and was not loyal to the state itself. All of this translated into a political environment which was far more highly charged and in which the stakes were much higher than in the relatively benign and gentle atmosphere of post-independence Prague. The ongoing transition to liberal democracy was compromised and held up as a direct consequence of the split. The space was created for and filled by demagoguery and populism. Economic change became intertwined with the political polarisation which resulted. Slovakia looked, for a while, as though it had not so much been set free as let loose.

With this in mind, the chapters of this book do not treat the Czech and Slovak republics as though they were really still part of the same Czechoslovak entity which gave rise to them. Broadly speaking, the four chapters focusing on the Czech Republic look at issues of relevance to a post-communist country which has come to terms with itself as an independent state, and the two looking at Slovakia discuss one which has not. The four to two division may seem unfair. But it takes into account the fact that the problems of the two countries are of an entirely different order. Also, the sense in which issues of universal importance can be teased out of the respective countries is not identical. Discussion of Václav Havel, for instance, can instruct us on the dangers, difficulties and rewards of an intellectual involving himself in the dirty business of high politics. With no comparable figure available from the Slovak political scene, there seems little point in forcing out a chapter on Michal Kováč, the Slovak president for several years, whose significance to his country was enormous, but for entirely different reasons. Likewise with the economy. The Czech privatisation process was a flagship for east European reform. Its successes and failings can tell us a lot about capitalism in general and the task of building a private property based economy in particular. Slovak privatisation, which started in the federation, became intimately connected with the style of government associated with the country's prime minister, Vladimír Mečiar. In general, combining discussion of Slovak political and economic transformation into one chapter explaining how the state came about, and one other analysing developments from that time forward, reflects a recognition that the core issue in Slovak politics for most of the decade has been Mečiar and Mečiarism. In understanding how this political personality was formed we simultaneously gain insight into the way in which the parties, civil society and the economy have developed in response. Czech Prime Minster Václav Klaus has come in for his fair share of criticism but his style of politics has not impacted on the Czech state in a remotely comparable way. Straightforward comparative analysis has its advantages but it can also lead us away from clear discussion of important issues.

Anyone inclined to suppose from what has been said thus far that the Czech Republic is about to be portrayed as a success story, while Slovakia is up for yet another beating at the hands of a foreign observer, is in for a surprise.

There was a time, in the early to mid-1990s especially, when it was almost impossible to read anything negative about the Czech Republic. The country's own leaders spoke openly about having reached the 'post-transformation phase' and foreign journalists, analysts and investors, perhaps taken in by the charm of the country's outstandingly beautiful capital, were inclined to agree. President Havel, a playwright and leading dissident, represented the quintessence of a head of state fitted to bringing his country back into the European mainstream. Prime Minister Klaus wooed foreign investors with uncompromising talk of the virtues of untrammelled market forces. One had brought communism down, the other promised to build capitalism up. This was a double act to be reckoned with. Ten years on, the heady enthusiasm of those early years has been matched by a sense of deep disappointment. In the latter part of the decade Havel was talking openly about a 'blbá nálada' or 'bad mood' having overtaken Czech society. Corruption appeared to have tainted the early ideals of the revolution. Political opportunism was rife. Finally, by 1998, the economy was spiralling back into deep recession. Klaus's economic miracle had been exposed as a sham.

Precisely in view of the high expectations of the early part of the decade we need to understand what want wrong. Everyone recognised that more than 40 years of hardline communist rule could not be erased immediately. But if there were many who said the transformation could not be achieved overnight, there were also those who seemed to suggest that the day after tomorrow might be a realistic alternative. The country's communist past turned out to be a much more challenging obstacle than many had supposed.

Criticism, nevertheless, should be kept in perspective. The Czech Republic faces serious problems but it is a secure democracy where the rule of law is respected and the rights of the individual are enshrined in a constitution inspired by liberal values. Likewise, there is consensus among parties of left and right that the state can do little to create wealth and that market solutions of one sort or another must be used to rejuvenate the economy. At the fundamental level, this is evidence of massive progress. To those who still want to argue that the Czech Republic is no longer a post-communist country per se, that though the country has problems they are not to be understood at an essential level in terms of the Soviet dominated past, this kind of very broad perspective can therefore provide some comfort. But sustaining this argument is bought at the price of missing all the important subtleties. It is when we come closer to the subject and split it up into its separate parts that we see just how clearly the Czech Republic remains a country locked in transition.

In Chapters 1 and 2 we set the scene with the emergence of Czechoslovakia in 1918 and the imposition of communism in 1948. The purpose here is to outline some of the formative experiences that the Czech and Slovak peoples would carry with them into the communist and post-communist eras.

Chapter 3 opens the way into the transformation years with a discussion of Václav Havel and his contribution to Czech society since 1989. This highlights Havel's devastatingly incisive criticisms of the communist system particularly in so far as he describes its core failings as a magnification of the problems of modernity as a whole. Armed with a critical apparatus of this sophistication Havel did not simply run out of intellectual energy the moment his maximal political aim had been achieved in 1989. It also seemed useful to begin with Havel because he carried this rich framework with him into battle against one time allies in post-communist society whose motives he quickly learned to distrust. Outlining his landmark speech at the Rudolfinum in 1997 (in which he describes what he thinks has gone wrong in the transformation process to date) sets us up for a more detailed analysis in the chapters that follow. It is the pivotal chapter of the book, swinging our attention backwards into an interpretation of the communist past and forwards into an indictment of the major failings of the present.

Chapter 4 takes us inside the mechanics of political transformation through the evolution of the party system and the way in which this should be understood in terms of a country emerging from communism. We also see how the development of party politics has brought into question the very morality-based presidency Havel had sought to establish. Chapter 5 looks at the economic transformation. The key argument put forward is that economic failure can be traced back to a deep conceptual flaw at the heart of Václav Klaus's understanding of what private property is and what was necessary to establish it as the basis for a capitalist system. Klaus's revolutionary ideas were formed in a communist environment where he learned to understand the advantages of capitalism only in their broadest formulation. Debating with those who opposed market economics as a whole he was never forced to engage the crucial subtleties. This deficiency left him poorly equipped in the monumental task of economic transformation which he conspicuously failed to see through. Chapter 6 discusses the patterns of behaviour suited to a changed political and economic environment. It is one thing to build a liberal democratic capitalist system, it is another to infuse it with the values and cultural substance which will produce the results expected of it. The style here is necessarily descriptive since the subject matter is the consciousness of a people. Even if opinion polls can give us some insight into the way in which the collective mind has evolved since 1989 any such findings are too vague for us to draw hard and fast conclusions. The intention in this chapter is rather to suggest which issues may have been important and thus get a general idea of how far the transformation process has come at the level of mass society.

In Chapter 7 we move to an analysis of the break-up of Czechoslovakia, examining institutional and structural explanations for the split but tying this back down into the developing national consciousness which eventually made federation unviable. Drawing on the notion of triple transition referred to above, we then move into the bizarre and worrying years of Vladimír Mečiar's rule over Slovakia, concluding with his defeat at the 1998 elections and an analysis of the country's prospects under a coalition government committed to redemocratisation and reintegration with the West.

Finally we look ahead to the future, focusing in on both countries' primary foreign policy aim of joining the European Union.

The book is neither a narrative nor a textbook. The emphasis is firmly on interpretation. No claim is made to have described every story or event of the last decade and the reader is assumed to have a certain knowledge of the subject. Developments in the former Czechoslovakia in terms of a particular communist past provide the thread running through the book and give it a unity of purpose, but the chapters can generally be read as single essays.

1
Building the State

Because most central and eastern European states are small in size and population and unprotected by high mountains or difficult seas, the fate of their people has always been determined by the good, indifferent or ill will of bigger and more powerful neighbours on either side.

One possible explanation for the way in which the identity of the region has been formed puts central and eastern Europe as a transitional area between East and West. In certain respects the validity of this observation seems obvious enough. With the exception of Hungary and Romania, all of the peoples of the area speak closely related Slavic languages while most use alphabetic forms recognisable in the countries of the West. Folk music and other national traditions draw from both sides of the continent. Their very centrality made the cities of the region into natural meeting points for merchants of East and West.

Less easily appreciable is the way in which geography may have influenced political consciousness in terms of the relationship between the individual and the state. Lying between the traditions of liberalism in the west and absolutism in the Czarist east and Ottoman south, it seems plausible to suggest, though impossible to prove, that central and eastern European nations fell prey to both influences, leaving a question open as to which way the cards would fall.[1]

But the clearest and most uncompromising consequence of being a small country in the middle of Europe was of course the reality of imperial domination and war.

Czechoslovakia emerged as a new state on the map of Europe in the aftermath of the First World War. What became known as the First

Republic was the product of domestic demands for autonomy backed by respected Czechs such as Tomáš G. Masaryk abroad, combined with an allied wish to find an alternative arrangement for the countries of central and eastern Europe under the control of the Austro-Hungarian Empire.

Czech and Slovak experiences of life under what were ostensibly the same imperial masters diverged widely and persist as aspects of the national character to this day.

Czechs, whose fate was determined in Vienna, were able to take advantage of a relatively benign and increasingly inclusive political and cultural atmosphere to gradually improve their lot. The Slovaks had an altogether tougher time. Ruled by Hungarians who saw Slovakia as little more than a province of their own country, linguistic, cultural and political rights were curtailed severely.

Despite protestations of a sell-out by some anti-Czech Slovak nationalists, therefore, the establishment of Czechoslovakia was at least as significant a step up for the Slovaks as it was for the Czechs although this does not mean they joined forces as absolute equals. An important part of the problem was that Slovaks had a far lower profile abroad than the Czechs. This not only affected Slovak perceptions of themselves, it gave Masaryk and his supporters a huge advantage in negotiations with the world powers on the form the new state would take.

Speaking a mutually comprehensible language, the two Slavic nations appeared to have a sound basis for a mutually beneficial cooperative arrangement, at least in the short term. Nevertheless, it is important to bear in mind one crucial point which would influence relations between the Czechs and the Slovaks throughout the coming decades. Although both had a strong initial interest in working together, a unity framed by opposition to the colonial past was bound to weaken as it became gradually clearer that the erstwhile imperial masters were simply not going to mount a comeback.

As such fears receded, the state would need new blood to sustain its vitality. It was one thing to establish a country called Czechoslovakia and another to get elite groups and ordinary people to believe in it.

On the popular level there were obvious problems. The Czechs and Slovaks certainly shared more linguistic and cultural similarities than either had had with their previous rulers. But Czechs spoke Czech and Slovaks spoke Slovak. Cultural development had taken place under different conditions and in the face of different obstacles. The Czech lands were more western in geography and the people, arguably, more western in outlook. Slovaks were devoutly Catholic. Czechs had a strong Hussite and also anti-clerical tradition. The major cultural threat to Prague had come from Germanisers, while Slovaks were rightly wary of their southern neighbours in Hungary.

There were of course similarities, which looked all the greater from afar, but was there enough in common to sustain the ties of allegiance necessary for a viable state in the longer term?

National distinctiveness therefore posed a problem from the very start. Success or failure in forging a specifically Czechoslovak identity would eventually determine whether the new republic had a future.[2]

If relations between the Czechs and the Slovaks were to be the cause of a long and painful headache for the country's various leaders, the presence of a vast German minority, mainly concentrated in the Czech lands, was a potential cancer threatening to overrun the whole body politic. At around three million, the number of Germans was not much different from the number of Slovaks. This represented a compelling practical reason for the Czechs to pull Slovakia into their ambit. With Slovaks in the fold, the mathematical balance quite simply tilted more sharply against the Germans and thus curtailed their influence on the character of the state.

The deal, however, worked both ways. The 1920 Treaty of Trianon, which carved up Hungary, had left more than 700,000 ethnic Hungarians concentrated heavily in Slovakia's southern border regions. Hungarian influence, and the dangers of irridentism, were diluted by the addition of the Czechs.

Both nations together, as Masaryk explicitly recognised, represented a bigger obstacle to the ambitions of each's minorities than either could have provided on their own.

But Czechoslovakia's Czech leaders had hopes which went far beyond a mere marriage of convenience and the terms of the relationship were made clear to the Slovaks from very early on.

There is no Slovak nation...the Czechs and Slovaks are brothers. Only cultural level separates them – the Czechs are more developed than the Slovaks, for the Magyars [Hungarians] held them in systematic unawareness. We are founding Slovak schools.

We must await the results; in one generation there will be no difference between the two branches of our national family.[3]

Masaryk's comments are fully representative of a generation of Czechs which regarded and sometimes even referred to the Slovaks as their 'little brothers'. The strongly Czech emphasis in Czechoslovakism was present from the beginning and would never entirely disappear, or be forgotten. When international conditions changed and after the Czechoslovak-enhancing problem of other large minority groups receded as an important issue, it was likely therefore that differences between the two major national groups would resurface.

Not, of course, that they were absent at the time. The unitary nature of the state dashed Slovak hopes for autonomy, a situation made all the more irksome by an overpreponderance of Czechs in the bureaucratic and financial institutions which governed national and economic life.[4]

In the context of the Great Depression, tensions arising from this rather patronising attitude to the Slovaks encouraged the development of ever more confident and avowedly nationalist forces in Slovakia, especially as the interwar years drew to a close.

The most visible expression of this tendency came from Father Andrej Hlinka's Slovak People's Party which eventually laid the foundations for Slovak 'independence' in March 1939. Slovak parties had been present under Hungary but they were small and lacked a mass voter base. Interestingly, and tellingly, there were no corresponding manifestations of Czech national identity in the form of Czech demands for greater autonomy. As rulers of Czechoslovakia – Slovakia was after all a mere appendage to the country's name – Czechs identified strongly with the Czechoslovak state which projected the Czech nation more powerfully as a force in the world. Slovaks may have been carried further on the international stage as a constituent of Czechoslovakia than they would have been on their own but they could not help but be seen as secondary.[5]

Despite Czech insensitivity – and it was a kind of snobbish insensitivity rather than an aggressive form of discrimination – Slovaks did achieve a good deal in the interwar years. The First Republic did not give them top dog status but it was a big improvement on the domineering rule of Slovakia's highly discriminatory, non-Slavic,

Hungarian masters in the days of empire. Slovaks could now participate at all levels of national life. Cultural organisations, Matica Slovenská being the prime example, were reborn. Literacy improved. Educational opportunities, especially at secondary school level, flowered, and enhanced freedom of expression allowed for increased use of the Slovak language in public life generally.

This may have been good news for the Slovaks but it was clearly a double edged sword for Czechoslovakism. At a time when policy was directed towards the formation of a collective national identity, the impact of this liberal cultural policy fell disproportionately on the national group which had been most repressed in the past. Slovaks were being asked to subsume their identity in Czechoslovakism at the time of their own nationality's most vigorous and free expression. At the mass level this was bound to be a source of confusion but it may also have affected the elite as well. After the disintegration of Austria-Hungary the many Hungarian officials running affairs in Slovakia retreated south to the motherland. Given the relatively poor educational opportunities afforded to Slovaks under Hungarian domination, the gaps that this left in the public administration were frequently filled by Czechs sent down from Prague. In the early years of the First Republic it may have been possible to sell this to the Slovak people as a price worth paying for the freedoms the new regime was providing. But this reasoning was gradually undermined by the very educational improvements noted above. By the second half of the 1930s a new wave of secondary school and university educated Slovaks had a personal as well as a national interest in taking over the positions once offered Czechs on grounds of necessity. Failure by the central authorities in Prague to move swiftly enough in recognition of this change not only angered the public, it presented a rising Slovak elite with first-hand experience of national discrimination.

Interwar Czechoslovakia may not have fully satisfied national aspirations but uniquely in the region it did allow for the reasonably free expression of national and political grievances. Relative to its time and geography it was a healthy democracy with most of the trappings of a free and open society. One factor militating against a deepening of the new state's democratic system was its very ethnic diversity which hampered the development of a unified public debate. Society was as much concerned with the rivalry of differentiated ethnic groups lying horizontally as with the vertical relationship between it and the state. Although political parties sprung up in their dozens the party system to some extent fragmented along national lines. This in itself contributed to the strength of a Czech dominated elite which remained in power for the whole of the inter-war period.

This problem aside, the political pluralism of the First Republic was a beacon of western bourgeois civilisation in a region seemingly dis-posed to authoritarianism. The structural conditions for success in this respect provided a solid foundation. The country was relatively wealthy. There was a substantial middle class lending itself to the development of liberal capitalist political groups. The high levels of industrialisation also provided the social base for a powerful Social Democratic party out of which a legal Communist Party was formed in 1921.[6] Apart from the party system, literacy and educa-tional standards were high. History worked in Czechoslovakia's favour too. The Czech aristocracy, partly as a consequence of the massive defeat suffered at the Battle of Bílá Hora in 1620, was too small to exert significant reactionary pressure. It was also blessed with a leader of outstandingly high calibre. Masaryk, a liberal sociologist and political philosopher, embodied the spirit of the Czechoslovak state. Like Václav Havel several decades later his personal credibility helped legitimise new institutions in a young democratic environ-ment.

Finally, Czechs under Austrian rule had been increasingly brought into the realm of public affairs. The Austro-Hungarian empire was not exactly the embodiment of autonomy and pluralism but the Czechs had sufficient experience of both to know what to do when called upon by history to manage their affairs on their own.

As Carol Skalnik Leff has suggested, this picture of a nascent Czechoslovak state striving to contain nationalist aspirations, depending for its freedom on events outside its own borders, confid-ent of itself, ruled by intellectuals and generally civilised in the ways of democratic behaviour shows remarkable similarities with the reborn Czechoslovakia in the first years after 1989. Czechoslovakia in the First Republic may have been a democracy but proportional representation led to a heavily fractured party political environment which constantly struggled to produce stable governments. In cir-cumstances of endlessly shifting coalition, policy was difficult to co-ordinate, a problem that was to resurface in the 1990s as well.

Another telling similarity noted by Leff was the uneven effect across the country of economic policy. One advantage Slovakia had enjoyed under Hungarian rule was a significant degree of protection for its industrial base. With this cushion removed, Slovak industry, itself relatively developed in regional terms, was exposed to the full force of competition from industry in Bohemia and Moravia. There was no evidence that Czechs were motivated by economically imperialist ambitions but Slovaks were bound to see their losses as Prague's gains. Not for the last time, Slovak national sensitivities were pricked by the unequal effects of an economic environment which appeared to hurt them while leaving the Czechs relatively unscathed.[7]

Domestic troubles aside, the most pressing problem the Czechoslovak state faced was of course from abroad. The country was surrounded on nearly all sides by countries resentful of territorial concessions they had been forced to make after the First World War. Foreign policy would not have been easy if Czechoslovakia had only had to contend with Hungarian discontentment. Germany, of course, represented the threat of potential and, later, actual, catastrophe.

Ethnic Germans (normally referred to as Sudetens) represented 3.1 million people out of a population of 13.9 million. In Bohemia the ratio was even more worrying. Worse still, the Sudetens were heavily concentrated in areas close to the border with Germany, a country which from 1933 had fallen under the ever tightening grip of a ferociously nationalistic government. Ethnic German demands would be increasingly encouraged by Nazi propaganda from next door and the Sudeten German Party became a powerful and threatening force on the Czechoslovak political scene.

Against a background of growing international instability, and the rapidly industrialising Soviet Union, the Munich Agreement of September 29, 1938, started a process that would end in a reality worse than the most pessimistic of worst case theorists could have envisaged.

Slovakia became 'independent' as a puppet state of the Third Reich, with Hungary gathering up some of its territory on the way. Poland struck out at Silesia. Finally, the Czechs themselves were subsumed under Germany as the Protectorate of Bohemia and Moravia. The West had sold Czechoslovakia out.

It could not have been evident at the time, but the Munich Agreement also tied Czechoslovakia's fortunes for the next half century to the Soviet Union. Stalin's failure to intervene on Czechoslovakia's behalf was seen as secondary, and probably attributable, to a western failure to show leadership against Hitler's aggression. Although it was going to be a very close run thing, Czechoslovakia was also in line for wartime liberation by the Red Army. Czechs had cause to thank Russia for its huge role in defeating Germany and this was not tempered by a feeling that the Soviet Union had been instrumental in their initial enslavement.[8]

In domestic terms, Czech perceptions of the two main 'minority' groups in Czechoslovakia – the Slovaks and the Germans – were bound to be changed by the experience of war.

The leader of the Slovak puppet state, Father Jozef Tiso, was brought to trial and subsequently executed. Significantly, Tiso was not only charged with collaboration, of which he was certainly guilty, but treason against the Czechoslovak state as well. Since Tiso had been charged with multiple offences the message was probably ambigous, but some Slovaks could not fail to draw the conclusion that Czechoslovakia's post-war leaders were making a statement about the place and the rights of Slovakia within the state.[9]

If the position of the Slovaks after the war was ambiguous, the status of the Sudeten Germans became brutally clear. No one in the aftermath of a war which had consumed tens of millions of lives was prepared to put sympathy for ethnic Germans anywhere other than bottom of their list of priorities. When the Czechs moved to exact their revenge nothing stood in their way.

It is difficult to overstate the critical importance of the Sudeten German question to any understanding of the character of the Czech people, their country and their national priorities. In the 18 months after May 1945, ordinary Czechs and then their government carried out the forced expulsion of nearly three million ethnic Germans from territories that their forefathers had lived on for generations.

The expulsions changed irrevocably the ethnic make-up of the whole state, but particularly the Czech lands of Bohemia which became something close to mono-cultural as a result. The national psyche of the Czech people was bound to be affected by the sheer scale of the operation.

It did not of course come out of a vacuum. The Nazi occupation of the Czech lands had been accompanied by thoroughgoing and utterly ruthless anti-Semitism combined with brutal reprisals against any Czechs who raised their hand against the regime.

The most notorious crime against the non-Jewish population followed the assassination of Reinhard Heydrich by Czech partisans in May 1942. The Germans responded in horrific fashion, burning the village of Lidice to the ground, executing all its 173 men and shipping the women and children to concentration camps.

Atrocities of this kind were less common than in some other countries in eastern Europe, but they could only harden popular sentiment against Germans of all descriptions. The Sudetens as a people quickly acquired the status of collaborators representing a fifth column inside the Czechoslovak state.

The forced expulsions began as a disorderly and largely popular expression of national vengeance but soon received official blessing under the terms of decrees issued by President Beneš.

Three million people did not give up their homes, their livelihoods and everything they owned lightly. Men, women and children were beaten, tortured and murdered in their thousands. Old people who could barely walk were kicked into line in forced marches where the sick and infirm died en route. Corpses were left to rot. Brutality was practised on a massive scale. This was real ethnic cleansing. The principle of collective guilt allowed for one of the most comprehensive attempts to purify the ethnic composition of a state in post-war Europe.

The communists had taken a leading role in the deportations but they did not need their pervasive propaganda apparatus to convince the population as a whole of the rectitude of what had been done. The people understood that instinctively.[10] The First Republic had shown itself incapable of sustaining itself as a coherent entity. It had failed the Darwinian test. But so had Germany and the Germans and they would be tried and sentenced by standards they themselves had established.

The consensus was not broken until the 1960s and 1970s when a angry but superbly insightful debate erupted among dissidents writing in samizdat.[11] The opening salvo was fired by Ján Mlynárik, a Slovak historian, who wrote under the pseudonym Danubius.

Mlynárik... wrote pointedly that Edvard Beneš's final decision to present the plan to the great powers on the expulsion of the

Germans was undoubtedly influenced by the 'motivation of the wider population, which wanted to make up for its own inactivity, if not collaboration, by identifying itself with the victors and by an ex post-"heroism" at the expense of the defenceless, which for the nation meant compensation, a release of opportunism, and bad conscience'.[12]

The sensitivity of Mlynárik's attack was all the greater because of his Slovak origins. To Czechs it seemed that the son of a collaborationist wartime Nazi puppet state was attempting to even up the score. As Abrams notes, they were angered that an outsider appeared to be insulting the national honour.

By the 1960s and 1970s few were defending the deportations on anything other than practical grounds. They were necessary to the stability of Czechoslovakia and the region as a whole. Those who cited human rights arguments were simply taking the whole affair out of its historical context. In any case, the war had shown beyond all doubt that territorial integrity and human rights considerations were sometimes inextricably linked. The Nazi's had implemented a reign of terror against the Czech people using the (generally willing) Sudeten Germans as the initial pretext. The argument had even been accepted and signed up to by the allies. Remove the Sudeten Germans and you had removed the underlying reason for the oppression of the Czech people. This was indeed fighting fire with fire, and the brutality was of course to be regretted. But 50 million people had just died in the most violent and bloody war in history. Drastic measures were required to ensure that it could never happen again.

The argument seems compelling enough, and it certainly so seemed to Czech society after the war. The revisionists on the other hand pointed up other consequences of the deportations. On the most obvious level, the Czech lands had deprived themselves of between a quarter and a third of their population. The Germans were often highly skilled and their absence left big gaps in the country's reserves of human capital.

But the crux of the revisionist argument was the effect the deportations had on the psyche of the Czech people themselves. By participating in such an enormous exercise in collective punishment the people had degraded themselves and compromised their instinctive commitment to law and morality. Policies stressing class responsibility and categorising entire groups as 'enemies of the state' would be that much easier to sell to the population at large once they themselves had bought so heavily into the collectivist mind set. Theft had also been practised on a massive scale. What had this done to respect for the notion of property rights? When the communists sought to nationalise the entire economy, what sort of defence would be mounted by a people who had just stolen the property of a third of its fellow citizens?

This line of argument is plausible but difficult to take beyond the level of enlightened speculation. Nationalisation, abuse of power, dictatorship, imprisonment, execution and brutalisation generally were the norm in all communist states in the 1950s, and it would be difficult to isolate the effect of the Czech expulsion of the Germans from the standard methods of Soviet style rule practised in the region as a whole.

Havel and the dissidents who emerged from obscurity to power in 1989 did not fail to air the issue in public soon after the revolution. But it took until February 1997 for the Czech lower House of Parliament to approve a reconciliation agreement with Germany in which Germany expressed sorrow for the wartime occupation and the Czechs expressed 'regret' for the expulsions. Nevertheless the Czechs pointedly refused to apologise and the issue of restitution of lost property is firmly off the agenda.

The affair touches on the integrity of the state and along with communism it represents the most significant factor in the development of Czech society in the last 50 years.

Czechoslovakia as it emerged from the Second World War had been through a whirlwind tour of its own potential and its corresponding limitations. Formed at the behest of the allies in 1918, it had been betrayed by many of the same countries 20 years later. Created out of the hope for Czechoslovak unity, it had been dissolved by a Nazi regime which enlisted its massive German minority and its fratricidal Slovak partner against it. Democracy, pluralism and the liberal-bourgeois values of the country's leaders had been unable to promote unity at home or face down hostility abroad. As a country in the centre of Europe, Czechoslovakia had given western values more than their fair share of attention. Perhaps it was ready for something from the East.

2
The Communists Take Power

Primed by the experience of two world wars in less than three decades and the economic depression of the 1930s, the post-Second World War consciousness of Europe was ready for change. The left was particularly well placed to take advantage. The Soviet Union had endured huge sacrifices in defeating the Nazis and this, combined with the leading role played by communists in the resistance movements, appeared to give radical groups the moral right to a certain respect. The fact that the Nazi–Soviet non-aggression treaty[1] in August 1939 had led directly to Hitler's invasion of Poland – the event which triggered the war itself – had been lost amid the mass of upheavals which followed. In any case, the mainstream European left had done such a comprehensive job of white-washing Stalin's monstrous crimes that the masses were largely ignorant of what communism in practice meant.

The decisive event leading up to the communist assumption of power in Czechoslovakia in 1948 was the general election which took place two years before. In the country as a whole the communists emerged as clear winners, though significantly they were beaten into a poor second place in Slovakia, leading some Slovak nationalists to complain later that, uniquely in Europe, their country had had communism imposed on it from the west.

The presence of a large industrial base, especially in the Czech lands, provided fertile ground for communist propaganda whose effective circulation was facilitated by the relatively liberal political conditions prevailing in the First Republic. As a legal political party the Communist Party had had time to establish a presence and agitate among the masses, a task made easier by widespread literacy and high educational standards.[2] The factors mentioned above, which lent credibility to left wing parties generally, had redoubled significance for a country which had been abandoned by the western powers at Munich. As we saw at the end of the last chapter, Czechoslovakia's bourgeois democracy had collapsed from within even as it was simultaneously disowned from outside.

With all this in mind it is not altogether shocking that voters in Czechoslovakia flocked to the hard left, giving communism one of its strongest ever democratic mandates in a free election. People did not of course vote for dictatorship in 1948, by which time communist support was falling sharply.

The timing of the 1948 takeover was neither merely fortuitous nor simply the result of a gradual consolidation of power by the communists since the end of the Second World War.[3] The Winter summit at Yalta in 1945 had divided Europe into spheres of influence within which the Czechs, so nearly liberated by the Americans, fell under Soviet control. This did not immediately translate into the imposition of the Soviet model of totalitarian rule. Stalin was not yet ready for confrontation with the West and with left wing parties enjoying a measure of popularity throughout the region he did not initially need to take the risk. Instead he secured the loyalty of his new possessions through what Hugh Seton-Watson described as a period of increasingly 'bogus coalition'. Communists took significant but not total control of the apparatus of state power, being sure to get their hands on the police and security forces, all the while building party support and laying plans for an eventual takeover.

The prelude to the coup in Czechoslovakia was the resignation of non-communist ministers precisely in protest over communist moves to take control of the police. The aim of the non-communist parties was to bring the government down and convene early elections in which it was hoped, reasonably given the opinion polls, that the people would vote for change. The plan failed partly because left leaning social democrats, who had not been properly consulted, were prepared to offer support to the communists, and also because the latter were less and less amenable to traditional political methods of being shamed into concessions.

With the West having already pushed a new charm offensive in the form of Marshall Aid, the Soviet Union had to move quickly or risk losing the gains it had won after the war. The May elections were conducted only among candidates approved by the communists. Democracy was dead and non-communists would be excluded from power for the next 40 years.

At the institutional level, the experience of communism in Czechoslovakia mirrored practice in the Soviet Union. Just as real federation within the USSR was rendered inoperative by the undisputed dominance of communist parties subordinated to central control from Moscow, the independence of states within the Soviet bloc was similarly compromised.

It is important, however, to recognise that an apparently similar model was mapped on to diverging national traditions in Eastern Europe to produce a different flavour in each of the USSR's satellites. This showed itself especially in terms of the varying attitudes to dissent and the forms of political repression generally.

With this in mind, the ill considered but widespread use of the term Stalinist to describe the early phase of communist rule in the East European countries needs some explanation.

Stalin's rule in Russia was characterised by waves of mass executions amounting in scale to genocide, the establishment of a vast labour camp system consuming tens of millions of victims, an atmosphere of mass terror throughout society and a cult of the leader's personality as complete as anything seen in history before or since. It entailed huge social dislocation in the from of industrialisation and an agricultural collectivisation programme which itself caused the deaths of up to 15 million people.[4] Stalin waged war on his own people and society was mobilised as a totality in the pursuit of his aims.

One can of course define terms as one wishes, but it hardly seems appropriate to use the same shorthand describing life under Gottwald in Czechoslovakia as for the 25 year reign of mass terror practised by Stalin himself in the Soviet Union. It is no belittling of the suffering of Czechs and Slovaks in the 1950s to point out that this was a difference of kind rather than degree. Some of the *trappings* of the Stalinist system were of course reproduced in Czechoslovakia. There was a state owned media mobilising society on the basis of a single dominant ideology, a state security police monitoring, harassing, imprisoning and killing opponents, a nationalised economy and the infusion, and therefore destruction, of civil society by the structures and subsidiaries of the communist party.

But in sharp and essential contrast to the Soviet Union under Stalin, Czechoslovakia was never, at any time, a society run along totally terroristic lines.[5] It was bad enough none the less. Political opposition was dealt with severely. In the 1950s tens of thousands were sent to prisons and work camps where many died, often under torture. Show trials rammed home the message that dissent of any kind would not be tolerated. Religion was attacked. Priests were arrested, harassed and vilified. Seminaries were closed down and believers faced ostracism and exclusion from social advancement for themselves and their families.

In the years following 1948 Czechoslovakia's communist leaders squandered the country's place as one of the wealthiest in the world, beat the population into a regimented submissiveness, poured class hatred into their homes through official propaganda, taught people to inform on their friends, live a life of dishonesty, cultural and spiritual decay, and destroyed large parts of the environment to boot. Deprived of the possibility of travel to the West, Czechoslovakia, which had already lost its cosmopolitan character through the murder of the Jews in the war and the deportations of the Germans which followed it, became an insular and, relative to its own past, culturally impoverished country.

In common with other east European states, the economy was subjected to the dubious rigours of central planning and heavy industry was given priority over Czechoslovakia's other relative advantages in light manufacturing and semi finished goods. Trade was redirected to the east and agriculture collectivised along Soviet lines.

The country also came under the sway of a foreign power, becoming part of the last great empire in Europe. Apart from its obvious construction along ideological lines, the Soviet empire had at least one highly unusual characteristic in that many of its colonies on the periphery were wealthier and more advanced than its core. For no country did this have greater significance than Czechoslovakia, which had not only avoided massive destruction in the war but also had the means to buy off popular discontent at repressive political practices through a relatively high standard of living.

This may partly explain why, in contrast with some other countries in the region, the partial denunciation of Stalin by Khrushchev at the Twentieth Congress of the Communist Party of the USSR in 1956 had little immediate effect on Czechoslovakia's leadership. Popular discontent on the back of social deprivation was largely absent as a factor driving the leadership into concessions.

Nevertheless, the country's reserves of wealth would not last forever. The inefficiencies of communist economic policy would, at some point, begin to bite.

The Prague Spring

In contrast with Hungary in 1956, pressure for reform in Czechoslovakia mainly came from within the communist party, although it was subsequently bolstered by society at large.

The man who was to emerge as the key player, Alexander Dubček, moved away from his native Slovakia at the beginning of the 1960s to join the secretariat of the Central Committee. His brief was industry. In his new position, Dubček was brought face to face with the country's industrial decline. Given Czechoslovakia's success as an industrial power up until 1938 – among Hitler's greatest early gains were the Škoda arms and engineering factories – Dubček and others in the party could hardly fail to start contemplating the idea of a mismatch between Soviet style economic planning and the needs of his own country.

Among many parallels with Mikhail Gorbachev's development as a reformer in the Soviet Union, Dubček initially understood the root of the problem as a failure to fully implement policy directives. If discipline were reimposed, central planning would work in the way that the party expected. Dubček came to the wrong conclusion but he had at least acknowledged that there was a problem. A more sophisticated analysis came from an economist, Ota Šik, who joined forces with the government to tackle the economic problems head on.

Šik's radical plans built on ideas emanating out of Khruschev's Soviet Union and drew subsequent inspiration from Soviet Premier Kosygin's reform efforts which were unveiled in 1965. He argued for decentralisation of decision making, a reduction in price controls, and the importance of market forces. This on its own entailed profoundly political consequences.

In Soviet type systems of government the party apparatus was charged with control or supervision of all aspects of life.[6] Private property and markets, therefore, are not merely disfavoured by totalitarian systems; they represent their antithesis. Market decisions are decisions arrived at by freely consenting individuals. Private property represents a space, controlled by the individual, which cannot be trespassed upon by others, government and party included. The issue was not merely a matter of dogmatics. Hundreds of thousands of party members depended for their livelihoods, and the privileges that went with them, on running pricing committees, planning offices, state enterprises and so on. The play off between market forces and totalitarianism was a zero sum game. The prospect of deep seated market oriented economic reform in all unreconstructed Soviet bloc countries had powerful political implications both at the level of ideology and at the day to day level of security for the established elite.[7]

With these considerations in mind, party leader Antonín Novotný largely shunned Šik's recommendations. But the genie was out of the bottle. Šik's ideas had been aired publicly and significant sections of the party had been impressed.

The genesis of the Prague Spring can partly be traced back to the second wave of 'destalinisation' in 1961. Thousands of political prisoners were released and many former communists, including the Slovak 'bourgeois nationalists' such as Gustáv Husák, stepped out of jail and eventually back into the party. Outside the party, artistically minded intellectuals were pushing for a relaxation of cultural censorship. Emboldened by events in the Soviet Union, writers and filmmakers including Milan Kundera, Václav Havel and Miloš Forman achieved national and international celebrity.

Added to widespread concerns within the elite about the stagnating economy, Czechoslovakia had the critical mass of discontent to give reformist tendencies a real chance of making their presence felt.

Party political factors also worked in the reformists' favour. Novotný's clumsy handling of the Slovak national question split hardline opposition at precisely the point when it needed to be united.[8] The final and crucially important piece of the jigsaw puzzle, sealing Novotný's fate, was the removal of Soviet support which may have come in response to public criticism voiced by Novotný some years before at the removal of Khruschev in the Soviet Union.

The coalition of forces hostile to Novotný did not, however, translate directly into unqualified support for Dubček. Managers wanting more economic power for themselves combined with apparatchiks looking for a more liberal interpretation of centralist principles, federalist minded Slovaks, and those who were simply out to get rid of Novotný for personal reasons.

Outside the party, the cultural intellectuals, of course, had an agenda of their own. Dubček's elevation to power in January 1968 may have drawn a veil over Novotný but it was by no means certain what would come next.

The bombshell which Dubček and his coterie eventually dropped came with the 'Action Programme' adopted at the plenary session of the Central Committee on April 5, 1968 and published in *Rudé Právo* five days later. Dubček referred to it as the 'heartbeat of Prague Spring'. Its very formulation, drawing on rank and file members and local party organs, marked a sharp change from the past. This was nothing compared to the content.[9] The document begins with the standard assessment of the rise to power of the communist party. As Dubček said, the document could not be a provocation. 'Czech and Slovak reformers were not out of their minds.'[10] The tone was cautious and the heretical nature of the content was interspersed with warnings about the 'aggressive attempts of world imperialism' and thus disguised by the traditional references to Marxist-Leninist ideology. It upheld the idea of the leading role of the party and it pledged Czechoslovakia's allegiance to the Warsaw Pact. But it also made some of the most radical demands for democratisation ever to come out of a Soviet Bloc country.

> A more profound democracy and greater measure of civic freedoms will help socialism prove its superiority over limited bourgeois democracy and make it an attractive example for progressive movements even in industrially advanced countries with democratic traditions.[11]

With this aim in mind the Action Programme called for the abolition of censorship, the systematic use of opinion polls in the decision making process, the inclusion of a multiplicity of groups and organisations in the administration of the state, open debate with representatives of 'bourgeois ideology', the right to travel abroad, further rehabilitation of the unjustly persecuted, the replacement of incompetent officials, changes to the electoral system, devolution of power in the party, the introduction of a socialist market economy and greater autonomy for Slovakia in the form of a new federal arrangement. In short, it offered the prospect of socialism with a human face.

With a set of policy initiatives of this magnitude, the dangers to traditional Soviet style rule were obvious both to substantial sections of the party at home and, crucially, to other communist regimes abroad. The Soviet leadership could hardly fail to be concerned.

This proposal, I should say, was immediately viewed by the Soviets as the beginning of a return to capitalism. Brezhnev made this accusation directly during one of our conversations in the coming months. I responded that we needed a private sector to improve the market situation and make people's lives easier. Brezhnev immediately snapped at me, 'Small craftsmen? We know about that! Your Mr. Bat'a used to be a little shoemaker, too, until he started up a factory!'

But the main objection to Dubček's initiative was political. When Dubček said he wanted to retain the party's leading role he meant it literally. The party would lead but there would be other groups coming from behind who could challenge its dominance. This was dynamite for the system as it had previously existed. Marxist-Leninist theory had always given an assurance that history itself had granted the communist party the right to undisputed dominance.[12]

Society took its cue. Independent groups attracted mass membership particularly in the universities, where the official youth organisation was superseded by the Union of University Students centred in the Philosophy Faculty of the Charles University.

Daringly enough, the union published its own Action Programme infused with the ideals of the United Nations Declaration of Human Rights. A non-aligned political grouping, KAN, pledged to nominate candidates to run in the next elections. Circulation of literary publications skyrocketed and groups of former political prisoners came together to form a pressure group, K-231 (the legal article under which they had been charged), to push for a review of their cases. In short, Czechoslovakia was developing the civil society foundations of an emerging, pluralistic, western style state.

Dubček knew how nervous the Soviet leaders were. But as his memoirs testify he was taken completely by surprise at the August invasion. His misreading of events was such that he believed military action to be 'unthinkable'. A series of meetings and conversations with Brezhnev and other Soviet bloc leaders throughout the year had not led him to take widespread rumours of military intervention seriously. Part of the problem, Dubček said, was the ambiguous language in which communist leaders expressed themselves. Words such as democracy and sovereignty were commonplace in the newspeak language in which such people worked, but they were either empty of any real meaning or meant the opposite of what most people thought them to mean. The only matter on which the Soviet leaders were unambiguous was adherence to the Warsaw Pact, which Dubček had no intention of leaving anyway.

Nevertheless, the 1956 invasion of Hungary should have served as an important warning. Dubček was, as he admitted, simply naive. But his description of this as a naivety 'rooted in goodwill and common decency' is too kind an assessment. More accurately it was borne of dishonesty. Like all communists of his time, his critical faculties had been warped by the lifelong need to lie to himself and others about the true nature of Soviet communism. The foundation of Lenin's state in November 1917 was hailed as a victory for democracy despite Constituent Assembly election results shortly afterwards which showed the Bolsheviks in second place to the Socialist Revolutionary Party. Lenin had abolished all other political parties, established a secret police force, used mass terror and set up concentration camps.[13] Stalin had murdered millions and yet it was communism of Stalin's type that attracted Dubček into politics. (He had actually lived in Russia under Stalin's rule.) Dubček had been a member of the Czechoslovak communist party when Slánsky was murdered and when tens of thousands were thrown into jail. The lies he had told himself to enable him to stomach crimes of this magnitude, which by his membership of the communist party he had given his tacit approval to, were bound to affect his judgement. Dubček said of the 1968 invasion that he simply did not believe the Soviet Union would do such a thing. He felt betrayed. The truth was that he had not allowed himself to see Soviet communist rule for what it was. Pluralism, democratisation, market reform and the abolition of censorship in Czechoslovakia not only represented the antitheses of Soviet style rule at home, they posed a real threat to the stability of other Communist states in the region whose people would undoubtedly be encouraged to ask for the same. Anyone who truly acknowledged the totalitarian essence of Soviet communism would have realised this. Dubček was wilfully blind to the system he had grown up around.

Normalisation and dissent

The result of the invasion was the implementation of one of the most restrictive totalitarian systems in the Soviet bloc. Normalisation, which to those inspired by the hopes of the Prague Spring meant precisely the opposite, forced all significant reformist forces out of the party. Every member was thoroughly screened. Around one-third either left of their own accord or were expelled. People were forced to denounce the reforms, which apart from the federalisation of the state were reversed. Dubček himself, having been kidnapped at gunpoint and taken to the Kremlin, was eventually shunted off to Bratislava as a forestry official. Many of the brightest and best simply left the country.

The idea of reform from within the communist scheme of things was also dealt a crushing blow. In society as a whole, the notion that socialism could provide solutions to its own problems had been dashed for ever. Dissidence would now explore the failings of the system as a whole rather than look for solutions from within. Liberal democratic, humanistic ideals were destined for a small, secret but effective revival.

In Czechoslovakia and throughout the region it had been made brutally clear that the eastern European states were little more than colonies of the Soviet Empire. Change internally would depend on change in Moscow. But the Prague Spring and its repression also affected the prospects for reform in the Soviet Union itself.[14] Soviet leaders had been given warning of where radical self-criticism could lead, putting them on their guard against the manifestation of any such tendencies at home.

Dubček was replaced by Gustáv Husák. From the Soviet point of view, it was perhaps believed that as a Slovak he would appeal to his fellow countrymen back home, and as a former political prisoner it may have been thought a useful symbolic gesture that reform was not entirely dead. Any such hopes were to prove short lived. Husák proved to be as slavish a 'normaliser' as the Soviet leadership could have wished for.

The conditions under which Czechoslovak dissent was expressed in the 1970s and 1980s were tightly constrained but not extremely violent. In one sense this was a product of the totalitarian nature of the system. The overarching control by the communist party meant that dissidence could be contained by subtler forms of coercion. The threat of job losses or non-promotion was one simple way of keeping ordinary people in line. Threats to children's prospects of going to university also blackmailed many into passivity. Conversely, joining the communist party could secure access to high positions and possibilities to travel abroad.[15]

More generally, what was the point of opposing the regime? The invasion had shown everyone that the country's leaders could count upon the might of the Warsaw Pact's military machine to deal with any substantial opposition. The massive informer network, controlled by a pervasive secret police force, made it impossible to mount any form of sustained political protest. In the end the jails could be and were used to silence the most dangerous critics. But there was no need to resort to mass terror to maintain order. The Czechoslovak government waged a kind of cold war against its own people. Few had the courage to stand up and fewer still knew what to do.

They were few indeed, but they were there. Some of the people in the dissident movement (Havel is justly celebrated as an outstanding example) showed courage and above all probity of the highest order in their principled refusal to bow to the enormous pressure directed at them by the full force of a totalitarian state. The dissidents were not a political party. They were, mostly, not political in the normal sense of the word at all, although in the context of a totalitarian state they were inevitably viewed as such by the party.

Many were writers, literary critics, and almost all were intellectuals of one form or another. They were motivated by what Havel described as a commitment simply to live in truth when all around they were surrounded by lies. The dissident movement, embodied by Havel, was determined that although it could do little to remove the structures of communist power, it could oppose them by simply refusing to consent to the lies which sustained it.

Charter 77 was a mainly Czech organisation, grouping former communists, anti-communists and non-political intellectuals. In this respect it had nothing substantial in common with the mass dissent expressed through Solidarity in Poland. Its main aim was to get Czechoslovakia to adhere to its obligations under the 1975 Helsinki Final Act.

Its philosophy was outlined in January 1977 by spokesman Jan Patočka.

No society, no matter how good its technological foundation, can function without a moral foundation, without a conviction that has nothing to do with opportunism, circumstance, and expected advantage. Morality, however, does not exist just to allow society to function but simply to allow human beings to be human. Man does not define morality according to the caprice of his needs, wishes, tendencies, and cravings; it is morality that defines man. This conviction is voiced in Charter 77 which is the expression of the joy of the citizenry that their country, by signing a document confirming human rights, a signature that made this act a part of Czechoslovak law, avows the supreme moral foundation of all things political.[16]

The inspiration for its foundation was the 1976 trial of an underground rock band, 'The Plastic People of the Universe,' which Havel attended as an observer. In an essay in October of that year, he describes how the proceedings of the kangaroo court which conducted the trial marked a watershed in the post-1968 era.

Havel sees the events as a microcosm of life in post-Dubček Czechoslovakia. The trial had the trappings of normality. There was a prosecution and a defence. The judge listened to both sides with apparent disinterest. But against this there was an uneasy realisation that everything was fundamentally dishonest. A kind of double life was being played out in which words and actions were not what they seemed. Reality and unreality appeared to be fighting for control and neither could quite establish themselves. The process was a kind of 'slipping out of joint'.

Somewhere deep down, however, I discerned yet another element in this experience, perhaps the most important of all. It was something that aroused me, a challenge that was all the more urgent for being unintentional. It was the challenge of example. Suddenly, much of the wariness and caution that marks my behaviour seemed petty to me. I felt an increased revulsion toward all forms of guile, all attempts at painlessly worming one's way out of vital dilemmas. Suddenly I discovered in myself more determination in one direction, and more independence in another. Suddenly, I felt disgusted with a whole world, in which – as I realised then – I still have one foot: the world of emergency exits.[17]

From then on there was to be no way out. Havel and the other dissidents who founded Charter 77 and VONS, the Committee to Defend the Unjustly Prosecuted, were prepared to and frequently did go to jail rather than compromise with the lie that lay at the heart of the communist system. Their struggle was a lonely one. By the mid-1980s the dissidents may have numbered only around a thousand or so in a country of over 15 million. It wasn't as though they could have defeated the system in the material sense if there had been millions of them. But their significance was enormous. They testified to the refusal of the spirit of freedom and truth to lie down in the face of overwhelming odds. Their very existence was an indictment of the corrupt elite which ran the country and it pricked the consciousnesses of the population at large.

Russians to the rescue

For reasons of obvious historical parallel, Mikhail Gorbachev's twin policies of Glasnost (openness) and Perestroika (restructuring) were never likely to be welcomed by the communists who ran Czechoslovakia in the late 1980s. Those of a reformist bent had been thoroughly purged in the months and years following the 1968 invasion, and the lesson which Warsaw Pact tanks had rammed home to those who remained was that far reaching political and economic reform could only end in disaster. From a purely self interested point of view they were of course right, as events in the Soviet Union itself were to show.

For some, the year 1989 marked the end of history. Liberal democracy had won a historic battle: though the style of governance would differ according to historical and cultural conditions, its democratic, market economic, law based form would prevail everywhere. The battle of ideas had been decided. The cold war had been fought and the West had won.

Others saw things differently. The end of communism marked the rebirth of a history in eastern Europe that had been held in suspended animation since the Second World War.[18] The national tensions that had been held in check by the rigidities of the cold war would resurface and play themselves out, bringing new dangers to the region and the continent as a whole. Still another way of characterising the events of 1989 is to see them as the year in which Europe finally came of age, ending the great European civil war of 1914–45.[19] The Soviet bloc, enforcing the dissolution of the continent's pivotal power, Germany, was a decades long hangover from which Europe had finally recovered.

Ironically, the underlying cause of communism's eventual collapse, and the inspiration for the policies which engendered it, was the failure of the system's economic base. Communism, it seemed, contained the seeds of its own destruction.

By the early 1980s the Soviet Union had become a military superpower resting uneasily on the fragile back of a second world economy. The confidence which sustained the Soviet leadership had long ceased to be drawn from devout faith in the ideological tenets of Marxism-Leninism. These were trotted out with ceremonial regularity but few believe that the politburo searched for inspiration in the texts of Lenin let alone Marx. Ideology did perhaps act as a limiting factor, or as some put it a sieve, preventing certain policies from being actively pursued, but it was little more than that.

The thought sustaining the self-confidence of the Soviet leadership was the Soviet Union's status as a superpower. The world looked for Moscow's reaction to every major event. The Russians mattered. They had a sphere of influence. Along with the Americans they decided the fate of nations around the world. But they could do so only in proportion to their credibility as a viable competitor to the West. This depended on two factors; the strength of the challenge mounted against them and their own ability to produce the economic conditions on which their military power was founded. By the 1980s both of these forces were moving decisively against them. Personified in the figures of Ronald Reagan and Margaret Thatcher, the West had found a new confidence. Reagan and Thatcher no longer argued that capitalism was a kind of practical necessity but that it was a moral imperative. Communism, by the same token, wasn't a nice idea which didn't work in practice, it was a profoundly evil system which had to be countered, pushed back and ultimately destroyed.

The Soviets had never had to contend with quite this level of hostility. Every move they made was matched and exceeded. America and Britain showed they meant business in an arms build-up that the Soviets would have to equal. But by the early 1980s the USSR's economy had ground to a halt. The Soviet Union faced a new and provocative challenge from the West at precisely the time it could least afford to respond. It became increasingly obvious that radical economic reform was unavoidable.

And yet, as Gorbachev came to realise, this could not be effected without fundamental political change as well. The problem was partly that the underlying failing of the economy was its communist dominated, bureaucratic nature. Changing the form of economic decision making meant demoting those who had been making the decisions. Also, the increasingly high tech nature of a modern economy made the totalitarian system less and less appropriate. Information technology depended on information. And information was produced through debate, discussion, disagreement and international contact. Soviet physicists noted that whereas it had been increasingly obvious that most of the important scientific innovations had been coming out of the West for many years, by the 1980s Russian scientists were even having problems understanding the theoretical discussions in the western academic journals. With the growing sophistication of weaponry this had a direct impact on the Soviet Union's ability to compete in the arms race. The American Star Wars initiative may have been little more than a science fiction dream but its suggestion as a serious policy initiative undoubtedly exemplified the fact that East and West were playing in different ball parks.

There was a limit to how far economic reform could go without breaking even the vestigial ties of the ruling elite to Marxist-Leninist ideology. Ideology may not have been a guide to policy but it was crucial as the party's justification for monopoly rule.[20] But if Gorbachev did not allow the reforms to go far enough he would make no substantial progress in revivifying the economy and society. There was also no point of balance which could have allowed Chinese style market reforms to coexist with the maintenance of party control. The forces in Soviet society aligned in Gorbachev's favour were those demanding both political and economic reform and those who opposed him were those who wanted an end to both. His supporters were frequently to be found outside the party. For Gorbachev to successfully outmanoeuvre the hard-liners, he needed to broaden the political environment in order to dilute his opponents' strength. He also needed to discredit them by exposing and renouncing their methods. What started as a plan to reform the flagging economy had a hard internal logic which pushed events far beyond their progenitors' initial intentions. These events were, of course, being watched closely in eastern Europe by leaders and people alike.[21] Gorbachev's visit to Prague in 1987 opened up a clear divide between popular feelings of hope and party elite concerns that perestroika and glasnost could spell deep trouble.

When a Soviet government spokesman glibly described the difference between Gorbachev in the 1980s and Dubček's Prague Spring reforms of 1968 in terms of the '19 years' that had passed in the meantime, the Czechoslovak leadership should have been bracing itself for big problems ahead. When it became clear that the Brezhnev doctrine was finished and that neither the threat nor the reality of Soviet tanks could be relied upon to save them, nervousness among the top leadership turned to barely submerged panic.

They still, of course, had their own apparatus of state control. The secret police and the informer network had not collapsed because of Gorbachev. The jails and the jailers, the riot police and the army were all effective instruments of control. In short, Czechoslovakia still had the shell of a totalitarian state; what it lacked in the end was the internal energy to mobilise these forces in its own defence.[22]

Although the communist surrender of power in Czechoslovakia was one of the last in eastern Europe it was also one of the most complete. One reason for this was the timing. The opposition was emboldened to ask for more than the Poles, whose power sharing agreement in the Summer of 1989 represented a huge step forward at that time. The other can be traced back to 1968 and the party purges which followed the invasion. Reform communists in Hungary could at least offer the example of their own experiments with liberalisation in the 1980s. In Czechoslovakia, communists after Dubček were dull witted, heavy handed, slavishly orthodox and rightly distrusted. It was difficult to do business with them and they had shown little sign of interest in any case. The morality based nature of dissent in Czechoslovakia also made the opposition less willing to compromise.

The Velvet Revolution began with a student demonstration on 17 November 1989. It was ostensibly called in honour of a student victim of Nazism but it quickly developed into a political rally against the regime. No one died in the violent assault on the rally by riot police on Národní Třída, but rumours of a death heightened tensions and the beatings administered on the orders of the Prague Communist Party caused an explosion of mass fury.

Crucially, the opposition broadened to include the workers, whose temporary general strike on 27 November sent a clear message to the country's rulers. The communist revolution had not only been called in the name of the workers it had given them real incentives to be loyal. A working class background made social mobility a possibility where once it had been a handicap. Czechoslovakia's communist rulers had also drawn up an informal social contract with the workers, guaranteeing them steady prices, provision of the basic necessities, narrowed differentials with white collar employees and even second homes in the form of widespread access to country cottages.

The deal was a cynical one but it had worked. Once the workers could no longer be bought off, the country's leaders knew that their time had come.

People in general were no longer afraid, or at least thought they had so little to lose and so much to gain that they were prepared to take the risk. The Berlin Wall had been hacked to pieces, communist regimes had fallen across the region. The Russians would not invade.

What started as a series of gatherings in theatres became the locus for the opposition and the establishment of an organisation, Civic Forum, which within a matter of weeks would take power. The leader of the Forum was Havel, a man little known outside Prague but whose vitality came to symbolise the ideals of the revolution. As Havel's star rose, the communists were forced into a series of concessions. The notorious Article Four of the constitution, guaranteeing the Party's leading role, went on 29 November. An effort by the communists to offer a compromise whereby they would retain significant control of the government on 3 December was laughed out of existence. On 28 December, Alexander Dubček became chairman of the federal assembly and, a day later, led Havel out to become the first non-communist President for 41 years. Pending elections in the middle of 1990, the government was a mixture of non-communists and communists but it was clear who was running the show.

Nothing symbolises the atmosphere of those days more than the appearance of Alexander Dubček and Václav Havel on the balcony of the Melantrich publishing house in Wenceslas Square on 26 November.

In Dubček's own words:

The powerful roar of the crowd as we appeared still echoes in my ears. Several hundred thousand people stood there cheering. My thoughts ran back twenty-one years to that May Day parade of 1968: I could compare this moment with nothing else. And that for me at least, closed the circle of historic events that had started in October 1967, when I launched the revolt against Novotný. So many things had happened – times of hope, times of defeat, times of patient resistance. Now I was standing on this balcony, at the side of a Czech dissident almost a generation younger than I, and we both knew that the crowd down there was giving us the power to bring the cause of freedom to its final victory in our country.[23]

As Dubček hints, his role at the time was symbolic. He was yesterday's man. The future belonged to Havel. The latter had proved himself as a man of principle in the dark days after Dubček's reforms had crashed to the ground in failure. Success as a dissident was one thing. In the new society that he was about to lead, Havel would be judged by different criteria. Was he up to the job?

3
Havel – Power to the Powerless

By the end of the 1990s the Czech people were beginning to show as much weariness at their playwright president as he was so evidently feeling at the demands of his office. Opinion polls had moved against him,[1] with some showing a small majority believing he should step down because of the chronic ill health he had suffered since undergoing lung cancer surgery in 1996. Criticism of Havel was less and less taboo in the newspaper columns and among mainstream politicians, hinting that the moral stature which had always outweighed his formal powers was on the wane. In a media game of guilt by association his judgement had also been called into question over the behaviour of his new actress wife, Dagmar, whose own deep unpopularity appeared to be dragging the president down as well.[2]

More generally, Havel, as head of state since the fall of communism, could hardly remain unaffected by the hardening public mood of pessimism in the country. Ten years of reform had brought many benefits but expectations had been higher still. With unemployment on the rise and a general feeling that the economic transformation process had been a failure, the establishment as a whole was bound to be tarnished. 'Havel fatigue' had set in and it seemed only a question of time before the great man gave in to the pressure and recognised that his usefulness to the Czech political scene had expired.

The two questions which everyone asked were those which had dogged Havel since his move from the theatre hall and smoky-bar life of the dissident intellectual to the splendour of high politics in Prague Castle. With the end of communism did he still have anything useful to say? Was Havel the independent intellectual cut out for politics at all?

To answer these questions we need to go back into Havel's dissident past, trace some important elements of his thinking and relate them to developments in the 1990s. The two main documents up for scrutiny are: his essay, the 'Power of the Powerless' – representing the high point of his dissident writing; and his speech to both houses of parliament in December 1997 – by far the most outspoken, controversial and dramatic moment of his (Czech) presidency. The two have been chosen for their timing and their content. They represent and encapsulate the most important periods of his life and are broadly reflective of the thoughts most pressing to him at each of those junctures.

By the time Václav Havel was 12 years old his country had lived through the end of a liberal democratic republic, Nazi occupation, liberation by the Soviet Union, the deportation of a quarter to a third of its (Sudeten German) countrymen, the quasi-democracy of postwar coalition and a communist coup.

Born into a wealthy, entrepreneurial family in 1936, his background was both a bonus and a handicap later in life. Havel's father moved in intellectual circles and was friendly with some of the leading cultural figures of his day, many of whom visited the family home. But Havel's bourgeois background and his family connections to the First Republic establishment were turned against him after the communists took power in 1948. He was prevented from pursuing any formal higher education in the humanities and he dropped out of technical university after two years.

As a teenager and a young adult, Havel grew up in the most repressive period of communist rule. The newspapers he read were filled with anti-bourgeois invective. Banners decorated the streets with the words of Lenin and Marx. Culture was brought under party political control and show trials and mass arrests made political discourse the stuff of whispers.

Although Havel is best known as a playwright, and although theatre occupied the most productive years of his early adult life, his political writings, especially from the mid-1970s onwards, are of greater importance.

In this respect it is little exaggeration to say that as a political essayist he developed into the quintessential east European dissident. Having said this we will shortly see how misleading it would be to interpret Havel, at an essential level, as an anti-communist. This is not to say that he was not fiercely opposed to communism, or that communist oppression did not provide the context in which he wrote. It is to recognise, however, that his opposition to communism was based on failings he attributed to modern civilisation as a whole. For Havel, communism had magnified the dangers of modernity to monstrous proportions, but it had not created them out of nothing. Many of Havel's most important criticisms as a dissident in communist Czechoslovakia were applicable in some sense to western societies as well. Once his own country had cast off communist oppression and adopted the values of the liberal democratic West, his critical apparatus would not simply run out of steam.

Havel's most important political essay, 'The Power of the Powerless', was written in October 1978. It is perhaps the seminal theoretical work underpinning and describing east European dissidence under communism.[3] It is Havel's statement of who he is, what dissidence means, what he sees as worthwhile in the world and what he thinks trivial or threatening.

Havel starts by identifying a core difference between the communist and other forms of authoritarian regime – communism's ideological character. But true to his concern for the universal nature of human problems he immediately places this in a broader context.

> In an era when metaphysical and existential certainties are in a state of crisis, when people are losing their sense of what this world means, this ideology inevitably has a certain hypnotic charm. To wandering humankind it offers an immediately available home: all one has to do is accept it, and suddenly life becomes clear once more . . . [4]

Ideology is portrayed as a surrogate religion answering the eternal human need for certainty. Havel explains the practical workings of this ideology through his famous example of a communist era shop owner putting the slogan 'Workers of the World Unite!' in his front window along with the onions and carrots. What is he up to? It is unlikely that he cares whether the workers of the world unite or fall apart in disarray. He may not even read the words. But this does not mean his actions are devoid of meaning. They are, Havel tells us, ceremonial but they are not merely incidental.

> The slogan is really a sign, and as such it contains a subliminal but very definite message. Verbally it might be expressed this way: 'I,

the greengrocer XY, live here and I know what I must do. I behave in the manner expected of me. I can be depended upon and am beyond reproach. I am obedient and therefore I have the right to be left in peace'.[5]

The significance of ideology in the communist system is real because of its ability to infuse society with dishonesty. By making such an apparently tiny concession the grocer is selling a part of his individuality through participation in the lies which legitimise the system and normalise it. It is the very ideological nature of the system which saturates communist society in lies. In order to survive, the party does not require that people believe in its ideology – in the post-revolutionary stages the leaders themselves may well not believe in it either – but it requires them to act as though they did. By paying homage to ideological necessities, people acknowledge the rules of the game. Mass ideology is the mechanism by which the system forces each to participate and share in the guilt of all. By telling one small lie, Havel's grocer makes the small lies of others easier to tell. The people may even be so immune to the propagandist slogans they mouth as to view them as amusing and make of them the subject of private jokes. What is important, though, is that they accept their life within the ideology. 'For by this very fact, individuals confirm the system, fulfill the system, make the system, *are* the system.'[6]

It is precisely with this recognition that Havel identifies the power of the powerless dissident. By living in truth, refusing to accept or live within the lie, the dissident threatens to break the metaphysical chain. His subversive power is specific to the kind of authoritarian system in which he lives. The poet, the playwright, the artist is generally unimportant to, and to this extent less powerful in, the traditional authoritarian system. The stability of the military junta, ruling as it does by naked force, is threatened by weapons, political organisation and confrontation generally. The ideologically based system (which also reserves the right to use violence if other means fail) is concerned with controlling society at a much more fundamental level. It must, above all, get involved with those who command and control language, for it is language, and its use, which bind the ideological system of rule together. This explains its uniquely obsessive concern with intellectuals. They are the makers of metaphor, the builders and the demolition experts of linguistic meaning, the guardians of the public conscience. Suppose, for instance, that the greengrocer and his customers read an essay in the newspaper by a non-conformist intellectual which suggests that the phrase 'Workers of the world unite!' is in reality a cover for their enslavement by a dictatorial regime and that therefore anyone who utters this phrase is a fool who has become a lackey of those in power. Whether the argument is persuasive or not, the signalling function of the green-grocer's daily routine has changed. Now he is no longer saying, I am beyond reproach, or, I can be depended upon. He is also risking an open statement of his own guilt: I can be depended upon because I am a collaborator. I am beyond reproach, in so far as enslavement by a dictatorship is an acceptable thing. The grocer may still put up the sign to save his skin but he can no longer do so with a quiet con-science. His passive and stable life has been threatened for the first time and he is now confronted with alleged realities about his society that had not been present to him before.

Those intellectuals who will not conform thus pose a direct threat to the ideologically based regime, which has offered the ordinary man a deal – a quiet life in exchange for ideological obedience. But the grocer's peace of mind cannot be sustained if the meaning of the public discourse which he buys into is subject to criticism and change. This is why there is cultural censorship and why the focus of repressive actions against intellectuals occurs in proportion to the ideological ambitions of the regime.[7]

So far the argument appears pretty specific to communism and ideologically driven authoritarian regimes in general. But Havel again broadens his point.

Is it not true that the far reaching adaptability to living a lie and the effortless spread of social auto-totality have some connection with the general unwillingness of consumption oriented people to sacrifice some material certainties for the sake of their spiritual and moral integrity? With their willingness to surrender higher values when faced with the trivializing temptations of modern civilisation? With their vulnerability to the attractions of mass indifference? And in the end is not the grayness and the emptiness of life in the post-totalitarian system [Havel's expression for the communist system he was living in] only an inflated caricature of modern life in general? And do we not in fact stand (although in the external measures of civilisation, we are far behind) as a kind of warning to the West, revealing to its own latent tendencies?

Havel's conclusion on the nature of dissidence and the disposition of the dissident is that he must refuse to participate in the lie at the heart of the communist system and instead commit himself to 'living in truth'. When dissidents have been doing this for long enough it becomes natural for them to pursue their life of truthfulness inside 'parallel structures' such as unofficial magazines, private concerts, seminars and exhibitions.

Havel's understanding of the life of the dissident has been discussed many times before, but it is important to see how this is leading him to a more general philosophy which will stay with him and guide him beyond the communist system's demise.

The 'live in truth' injunction to reject the material and ritualistic temptations of mainstream lifestyle does not lead Havel towards an introverted form of passivity. The parallel structures of dissident life are not akin to a monastery, insulated from what goes on outside. Quite apart from his bravery (for which he paid with years in jail and decades of harassment) in defending the unjustly persecuted, Havel's frame of critical reference is always employed to point up the broader picture. The values he examines are universals not mere particulars of the communist society in which he lives. And the universal problems which he addresses are those of the technological society, auto-matism, consumerism, a lack of believable ideals and, his increasingly dominant concern, the spread of secularism. Havel describes com-munist society with its atheistic anti-religion, its atomisation of the individual and its prioritisation of machinery and industry as the extreme representation of these general characteristics of modernity. In a section which prophesies the nature of the task Havel was handed after the 1989 revolution, he says: 'a solution cannot be sought in some technological sleight of hand, that is, in some external proposal for change, or in a revolution that is merely philosophical, merely social, merely technological, or even merely political.'[8]

There is no evidence that Havel believes democracy of the western type is equipped to find answers to the questions which most worry him. It may, he concedes, be better at concealing their importance because of the fundamentally healthier nature of society. It allows greater space for discussion of the problems, but it has not found any solution. However the answers are to be found, from whichever 'God who alone can save us', the result that Havel is looking for is a reinvigoration of values such as trust, responsibility and love and the relocation of man in the universe of higher values from which his materialistic and technological obsessions have alienated him. Havel has too shrewd an understanding of what is at stake here to venture a glib answer to the hidden core question which will exercise his thinking for the rest of his life – what hope is their for man and his values in the uniquely secular, irreligious world in which we are coming to live? But in the parallel communities of the dissidents and the human rights activists he does presume to offer up an example of the kind of institutions which could nurture a more morally worthwhile citizenry in the meantime. In the cooperative ventures the dissidents undertake, and in the spirit in which they undertake them, Havel sees, without using the expression, the significance of an embryonic 'civil society' as a giver of meaning, a creator of value and a locus of genuine community.

The key point here is that 'The Power of the Powerless', arguably Havel's most important political critique of communism and the place of the dissident within it, is, at its deepest level, not really an essay about communists or dissidents at all. It is a critique of the human condition in the modern world. In communism, Havel sees the dangers of the contemporary world through a disturbingly powerful magnifying glass. But he does not identify something which is unique or absolutely exclusive to it.

On one level this recognition shows how remarkably penetrating Havel's critical faculties were. How easy it would have been to emulate the many western analysts of the communist system who fell prey to parochialism and saw the repression of the individual and the destruction of traditional values as something specific to life in the eastern bloc. On another level it shows us how well equipped Havel would be to address the priorities of politics and society after communism had fallen apart. From his own point of view this probably explains why he did not regard the presidency he took up in 1989 as a mere prize for the heroic work he had done in the past. Havel did not simply bow out in favour of younger men after a few months or even years because he felt he was possessed of an intellectual energy which could sustain him beyond the environment in which his thinking had matured.

The great turning point in Havel's life was of course the Velvet Revolution, which almost literally delivered him from the hands of his jailers to the post of President of the Republic. His most important early address to the nation was on New Year's Day 1990, less than a week after assuming office.

His opening pledge is, predictably, one of honesty, and he closes with a pathos befitting the atmosphere of the time, telling his fellow citizens simply that their government has returned to them. The content is dominated by his description of the country's primary problem as one of moral decay, the roots of which he locates in the way the old system taught people to say one thing and think another. He acknowledges that everyone participated at some level in the totalitarian system, though some of course bear more blame than others. But he sees in the events leading up to the Velvet Revolution the expression of an underlying strength in his people, drawing on the democratic traditions of Czechoslovakia's past. Looking to the future he calls for his country to follow the ideals of its founding father, Tomáš Masaryk, and base politics on morality.

Let us try, in a new time and a new way to restore this concept of politics. Let us teach ourselves and others that politics should be an expression of a desire to contribute to the happiness of the community rather than of a need to cheat or rape the community. Let us teach ourselves and others that politics can be not only the art of the possible, especially if 'the possible' includes the art of speculation, calculation, intrigue, secret deals and pragmatic manoeuvring, but that it can also be the art of the impossible, namely the art of improving ourselves and the world.[9]

Contrariwise he sets out the anti-vision:

Our main enemy today is our own worst nature; our indifference to the common good; vanity; personal ambition; selfishness and rivalry. The main struggle will have to be fought out on this field.

Nearly eight years later, speaking to a packed audience at the Rudolfínum in Prague, Havel delivered the most important speech of his presidency. How successful did he believe his country had been in avoiding the vices outlined in the warning just quoted above?

> The prevalent opinion is that it pays off in this country to lie and to steal; that many politicians and civil servants are corruptible; that political parties – though they all declare honest intentions in lofty words – are covertly manipulated by suspicious financial groupings. Many wonder why – after eight years of building a market economy – our economic performance leaves much to be desired, and even compels the government to patch together packages of austerity measures; why we choke in smog, when so much money is said to be spent on environment protection; why all prices, including rents and electricity tariffs, have to go up without a corresponding increase in pensions or other social welfare benefits; why they must fear for their safety when walking in the centres of our cities at night; why almost nothing is being built except banks, hotels and homes for the rich; etc. etc. An increasing number of people are disgusted by politics, which they hold responsible – and rightly so – for all these adverse developments.[10]

Building on this devastating polemic, Havel moves on to the why and the who. Partly it is the result of what he calls a 'post-communist morass', which left many people poorly equipped to handle the responsibility which accompanied their new freedoms. Havel compares this with a post-prison psychosis and warns his audience that this state of mind has not yet disappeared from Czech society. But the bulk of the blame cannot so easily be passed off on the old regime. Those who are responsible are those who are sitting in front of him, the political elite which has allowed the unscrupulous and the criminally inclined to rifle the country's economy while they sit idly and arrogantly by and let it happen. Havel does not mention Klaus by name but as the leader of Czech political and economic reform since the Velvet Revolution it is clear that he is the primary object of attack.[11]

And at the heart of his complaint is the corruption, the lawlessness and the indifference to moral principles which attended the centrepiece of Klaus's political lifework – large-scale privatisation. Showing a subtlety of understanding of what is necessary to make capitalism work that renders Klaus as bewildered as a lost child staring back at a distant parent, unsure of what to do and left with no easy option but to cry out, Havel in one speech cuts to the quick of all that has gone wrong in public policy making in the years since 1989.

I do not share the view held by some of you that the entire transformation started from the wrong foundations, was wrongly devised and wrongly directed. I would rather say that our problem lies in the very opposite: the transformation stopped half way, which is possibly the worst thing that could have happened to it.

Havel, as he goes on to say explicitly, has understood that formal privatisation – the mere dumping of state assets, as if off the back of a lorry on to a street – is not real privatisation at all. There is a difference between possession and ownership. Possession is the preserve of the strong, property is the product of legality. Clearly enforceable and actually enforced rules are, therefore, as Havel recognises, the *sine qua non* of real privatisation. Havel asks rhetorically:

But how can we expect the desired restructuring of companies, and of whole branches of our economy, when there are so few clear owners, and when so many of those who represent the owners see their role not as a task, mission or commitment but simply as an opportunity to transfer the entrusted money somewhere else and get out?

Havel has understood the significance of transparency of ownership and the enforcement of fiduciary duty to the functioning of a capitalist system, while the architect of the country's economic reform and the one who accuses him of ignorance has spent eight years 'transforming' his country towards capitalism with only a limited appreciation of what was necessary to make that happen.

But as in his criticisms of communism decades earlier, it is the effect that this environment has on the moral life of the individual and the subsequent worsening of that environment as immorality becomes self-perpetuating that most concern him.

Havel identifies the corruption that Klaus has presided over as the key failing in contemporary Czech society. The culture of the people, Havel believes, has been affected such that to respect the law is to put oneself at a disadvantage to the privileged whose wealth is so often born of lawlessness. From this rational calculation, the instinctive bonds of cooperative morality which bind society together are let loose. There is pessimism, bitterness, a return to affection for the past or a move towards more elemental solutions from nationalists and populists. The culture which sustains a capitalist system – respect for contract, trust, thrift, and honest work – depends on the representatives of the elite epitomising and enhancing just those values. If they do not, or as in the Czech case they appear to actively oppose them, the mirror in which society sees itself reflected flashes back a warped and distorted image. People do not know how they should behave. In emphasising the crucial importance of a strong cultural base to a flourishing capitalist environment, Havel is allying himself with the concerns of the best and most advanced of capitalism's supporters in the West. So what does Klaus, the self appointed guardian of the Czech Right, make of Havel's concerns?

> He [Havel] pointed to a vision of the world that is diametrically opposed to right-wing thinking, and showed how deep is his ignorance of the workings of the market economy and a free society eight years after the fall of communism, one day short of the eve of the eighth anniversary of the appointment of the first post-communist government in our country.[12]

We will see in Chapter 5, on the economy, how superficial was Klaus's understanding of the subtleties of private property and the workings of markets. Here we see him failing in spectacular fashion to understand the cultural prerequisites of capitalism as well.

The point of all this is not simply to show that Havel as president continued to represent the best of Czech traditions in social and moral commentary. Nor is it to show that in building an argument around economic developments Havel, in his weakest area, had a more profound and rounded understanding of his subject than Klaus in what is meant to be his strongest. Nor even that Klaus's vision of the world is superficial and one dimensional and Havel's is not. The point, however, is to show that Havel is no grey and insular intellectual. He has real feeling for the way theoretical failings can translate into concrete problems, and vice versa, at the most day to day level affecting the most ordinary people. He is a genuinely political intellectual with a profound understanding of what is important to a modern capitalist society.

The answer then to the first of our two questions on Havel's performance in the 1990s appears to be clear: Not only did he have something to say after the end of communism but he was better at saying it than anyone else in the establishment.

And yet, it is from this reply that we are led to the question which Havel and his supporters are most vexed by: Was the establishment the appropriate vantage point from which to launch the assault? This does not merely concern the unavoidably reflexive character that his criticisms appeared to take on. How much credibility could Havel expect to be given in denouncing a political elite of which he himself was the most enduring, if largely symbolic, representative? It also went to the heart of the role of the intellectual in politics and whether the twain can in fact meet.

The argument was made most explicitly by Timothy Garton Ash in an article in the *New York Review of Books* in 1995.[13] Garton Ash, in a spirit of friendly criticism of Havel, argued that the relationship between the politician and the intellectual would always be adversarial because of the differing standards of truth they both abide by. The party politician worked in the realm of half truth, a fact that was highlighted by the meaning of the word 'party' itself, implying something partial and one sided. The party, democratic or otherwise, has a line and the politician's duty must lie more with supporting that line than pursuing his thoughts wherever they may lead him. The independent intellectual has no such constraints and is bound only by the limits of his mental capacity. As Garton Ash recognises, the argument is somewhat harder to make for a non-party head of state. But even he will be duty bound to participate in decisions arrived at through the party process if only because he must sign them into law, sometimes against his will.[14] Another way of expressing the same point is to recognise the essentially individualistic nature of the intellectual task at the methodological level. It would be going too far to say that the intellectual should be unconcerned with the way his ideas are received. He would never put pen to paper or open his mouth if he followed that maxim too strictly. But in so far as the intellectual tailors the conclusions of his arguments to suit the beliefs or prejudices of the world outside, he has compromised the most fundamental of his aims. The independent intellectual's critical frame of reference therefore must be internal. He will not be oblivious to the world at large, but his own conscience must speak with a louder voice. This then raises the question of what are the circumstances most likely to help him in his task. Presumably, there are some kinds of environment which produce external temptations making it harder for the voice of independent conscience to be heard. This, the Garton Ash argument goes, is precisely the position Havel finds himself in. Does the argument stand up? For the party politician, the point appears to be well made, and applies the more forcefully the higher up the political ladder one climbs. The guiding principle of cabinet government, for instance, is collective responsibility, meaning that faced with a policy the minister does not agree with he must pretend that he does in fact support it, that is lie, or resign. This is the very antithesis of an environment conducive to the intellectual life. But Havel is not a party politician. He does not have to follow a governmental line and he is answerable only to himself in his office. Or is he? Havel may not be the head of a government armed with a policy agenda, but as president he is the head of a state with a legitimising set of myths. In Chapter 1 it was noted that along with the experience of communism itself, the expulsion of the three million strong German minority was among the most formative events in the development of the Czech psyche this century. Havel has not shirked this issue entirely in his role as president. He has expressed regret at the suffering caused by the deportations but has fallen short of issuing an apology and has shown no inclination to offer compensation or put matters to rights. Now it is possible that Havel would have reached similar conclusions if he had been facing the Sudeten German question as a disinterested intellectual. The point is, however, that the Czech President cannot view an issue as fundamental to the character of the state which he heads with disinterest at all. The demands of his office dictate that he come to a conclusion which preserves the integrity of that state. Whether Havel's viewpoint would have been the same in other circumstances is beside the point. On one of the most significant moral controversies of his country's recent past, Havel the intellectual is rendered impotent by Havel the head of state.

A more obvious problem is that as a non-party president Havel is duty bound to express himself in ways that do not place him too strongly in support of or opposition to the policies of any particular party. Suppose for example that Havel supports one important policy which is passionately supported by the Social Democrats but fiercely opposed by Klaus's Civic Democrats. Undoubtedly this has actually happened, and also in reverse, month after month since the early 1990s. But the non-party president can hardly let his views be clearly known. To do so would be to compromise his position. The party politician may, as Garton Ash says, work in half truths, whereas the intellectual seeks to work in whole truths. The non-party president, however, risks working in non-truths, fearing to express his views clearly and deliberately generalising to avoid being too heavily linked to particular policies or ideas. When he is good, the impartial head of state can make impressive contributions by pointing up universal values. But when he is bad he is inclined to waffle and fudge in order to conceal the reality of a purpose which contradicts the conditions of his office.

Havel was criticised over his 1997 speech, even by those who agreed with him, for not having spoken up earlier. Such criticism is in fact misplaced. He has been meditating on the importance of civil society, legality and fairness throughout his presidency. What he had never done was so directly to criticise one set of politicians, to link accusations of moral and administrative corruption directly to the outgoing Klaus government. The crowning point of the argument is of course that he did it when Klaus's government was in fact outgoing. At this brief intermission in the party political life of the country, Havel the unconstrained intellectual could finally make truthful, concrete criticism of particular people for particular events. But what kind of a 'living in truth' is that?

The problem appears to be insurmountable. Has not Havel found himself in the worst of all possible worlds? In a speech in 1998 he described the role of the politician not just in terms of being able to perceive things but to put them into practice.[15] As a ceremonial president Havel appears unable to either freely express his perceptions or put them into practice. He is neither effective politician nor free thinking intellectual.

But this is only the most pessimistic assessment. We have already seen how much Havel has to say of relevance to life in the post-communist era. Perhaps the limitations of his office have too often forced him into generalisation and too infrequently to focus on specifics. But in generalising on important moral principles his office has at least given him access to an audience that goes far beyond the small circle of intellectuals who would have heard him had he spent the last decade as a journalist inside a literary magazine. The latter occupation may have made for a purer, more focused Václav Havel but it is not easy to say whether Czech society as a whole would have been better off as a result. How many heads of state are as capable as Havel of bringing issues of moral and philosophical importance into the political debate? It is clear that Václav Klaus dislikes Havel for doing this. But is not Havel playing a crucial role for his people: offering them a different image of what politics is from the harder, drier and narrower world of Klaus and Klausianism? One way of putting the case for Havel is precisely to ask what the Czech Republic would have looked like in the 1990s, how it would have felt, had Klaus been president and not Havel? By broadening the question the point becomes even clearer. Why has it been so difficult to come forward with any credible candidate for Havel's replacement? From that very difficulty do we not come to the grudging conclusion that Havel may have managed to square the circle we have just drawn around him? As we will see in the coming chapters, matters are far from that simple. But as a provisional conclusion: the country's greatest asset deployed in its most effective setting? Perhaps.

4
Politics After Communism

As we saw in Chapter 1, peacetime Czechoslovakia in the interwar period was the most stable and genuinely functional liberal democracy in the region. After 1989, the Masaryk inheritance therefore served as a reminder to Czechoslovak, and above all Czech, citizens that the standards they now aspired to had once been achieved before. The ideals of liberal democratic government, the rule of law and pluralism did not exist as mere abstractions. In the consciousness of the people, the transition to democracy could be taken as a reassertion of the national heritage and not as something alien or imposed from outside.

Quite how this expresses itself in the realm of practical politics is difficult to say. Its importance can perhaps be gauged by considering the experience of other countries in the region – Slovakia, as we will see in Chapter 8, is a good example – where anti-democratic forces have drawn inspiration from mythical golden ages when leaders earned their people's loyalty through strong or absolutist rule from the top. As communism gave way to freedom, dictatorial inclinations and national pride could easily become enmeshed. For Czechs, the great era of national assertiveness coincided with the flourishing of liberal democracy. Authoritarianism and nationalism represented a bad fit.

A decade after the Velvet Revolution it seems reasonable to ask how far the Czech Republic has managed to translate this initial set of advantages into practice. Above all, does it still remain appropriate to refer to the Czech Republic as 'post-communist' at all, or do the country's political problems roughly correspond to those which also affect the more developed democracies of the West?

The battle ground for this debate is clear enough.

The pivotal period for Czech politics in the 1990s after the split with Slovakia came with the fall of Prime Minister Václav Klaus's centre-right coalition in late 1997 until the establishment of a social democratic alternative with support from Klaus the following summer. The rise of the moderately leftist social democrats at the 1996 elections had altered the balance of power by depriving the centre-right of a working majority, but late 1997 marked the first time the country had seen a government fall, and it represented the point at which all the main forces on the Czech political scene came out into the open.

President Havel used the full authority of his office to condemn important features of the transition to date; the long-standing but partly submerged hostility between Klaus and the president and between Klaus and the other centrist and rightist members of the government was made public; the corruption which had done so much to undermine public confidence in the reform process became a key issue in the party political debate; finally and partly underpinning all of this, the period coincided with the descent of the economy into serious recession, bringing into question the entire economic reform project and especially the credibility of those that had conducted it.

The crucial result of all of this was that the party political system temporarily broke down. It was necessary to appoint a non-party government headed by central bank governor Josef Tošovský, and it became clear that stable parliamentary majorities were going to be difficult to achieve for a long time to come. The relationship between the presidency and the party system also, and for connected reasons, reached crisis point.[1]

With so much going on during this period and so many factors coming together at once, we need to pull the main elements apart, examine them separately and then see why the parties, and simultaneously the party system and the president have found consolidation and cooperation so difficult. This is not merely an exercise in proving that the direction of Czech politics has been determined by a particular communist past, although it will become clear that this is so. Since it is now apparent that the specific problem of Czech politics is the failure of democratic elections to produce viable governments, the country is considering changes to the electoral system aimed at promoting greater political unity. Long term success or failure in this aim will to some extent depend on how policy makers interpret the underlying causes of division.

The fracturing of party politics

In the shape of Civic Forum, the Velvet Revolution was led by what political scientists have referred to as a 'conglomerate'[2] party bringing together a wide variety of people united only by their opposition to communism. Petr Uhl, for example, one of the most prominent and persecuted of the dissidents, was a Trotskyist. Václav Benda started off as a kind of Christian democrat with a good word or two to say about General Augusto Pinochet's anti-communist coup in 1973. Václav Klaus was a hard headed neo-liberal economist with little time for the many dissidents who provided the revolution with its distinctively intellectual and moral character.

These individuals and the groups which formed around them found themselves working together in highly unusual circumstances. The Civic Forum umbrella organisation was a genuine product of the early transition from communism, and as the unifying enemy image of the old regime receded into the past, the disparate groups within it quickly went their separate ways.

Let us look at how the party political system has emerged since that time.

The right

The Czech right has been dominated in the 1990s by Václav Klaus's Civic Democratic Party (ODS) and one of two smaller parties. Up until 1998 this baby brother role was played by the Civic Democratic Alliance (ODA). Since then the Freedom Union (US), formed from dissident voices within ODS, has taken its place.

Looking at the nature of politics in the region as a whole, George Schopflin, writing in the first years after the 1989 revolutions, has argued that the right actually faced greater difficulties in adjusting to the new conditions after communism than the left, which he acknowledges had just suffered a massive defeat.

Because there was no opportunity to modernize conservatism under communist rule, the conservative traditions to which right

wing or 'moderate' parties hark back are those appropriate to a pre-
modern polity and society, that of the 1930s, nationhood and
religion. The trouble is that, in the meantime, major changes
have taken place in these societies and the ideas of the 1930s are
barely appropriate to the situation, while nationhood and religion
have little to say about the distribution of power.[3]

However sensible an observation this may be in other parts of eastern Europe it has little if any validity for the three parties of the Czech
right – ODS, ODA and US – whose inspiration has been clearly drawn
from the ideological precepts of the New Right. Religious and
nationalistic forms of conservatism have rarely surfaced.[4]

Although the argument from atrophy does not seem to hold in this
case, the legacy of communist rule has had one absolutely decisive
effect on the Czech right in depriving it of a class base from which to
draw support. The bonds of loyalty, formed through self-interest on
the part of established rural and industrial groups and the aspirations
of the entrepreneurial class, could not develop in a communist system which had abolished private property and in which social mobility was largely contingent on collaboration with the ruling party.
Much of the social base of the left, on the other hand, remained
intact. After the revolution, left wing parties had a pool of natural
voters comprising those who had always relied on state social
support, industrial workers, and of course, rump communist sympathisers and former party members.

Lacking a comparable class base therefore the right had to look
elsewhere to establish itself as a powerful political force. In the country's relatively enlightened political atmosphere, populist demagoguery was a non-starter. A serious political challenge would require a
strong set of policy initiatives.

Neo-liberal ideology, as nowhere else in eastern Europe, became the
legitimising set of ideas binding together the Czech right and ODS.
ODA and the US have seemingly embraced it with equal enthusiasm.[5]

There are several possible explanations for this. In the negative
sense, it seems likely that long-standing anti-clerical traditions and
a liberal and permissive social environment made religious based
conservatism an unlikely option for a mass party. Moral majority
type politics is only possible if the majority actually believes in that
kind of morality and there is little evidence that this so in the Czech

case. Crude nationalism also appeared to have limited appeal among a people that wanted to join the European Union and had no territorial claims or significant diaspora abroad. Neither did the Czech Republic contain significant minorities from other countries. Czechs may be more or less openly hostile to Roma at the individual level, but as the mixed fortunes of the Republican Party suggest, they do not substantially reward those who play out such sentiments at the level of high politics.

With these doors closed to the right, neo-liberalism may have simply suggested itself by default. But there are also positive reasons why it should have succeeded in the Czech case. For one thing, the electorate and the intellectual elite have undergone certain experiences favourable to the development of New Right thinking. Uniquely in the Soviet bloc under communism they had experienced a short lived attempt in 1968 to provide an alternative from within the socialist scheme of things. The failure of the Prague Spring reforms may have served as an inspiration to look for an entirely new ideological framework, if for no other reason than because socialism appeared to have exhausted its possibilities.

In addition, the experience of 1968 was bound to have brought many Czechs to a deeper understanding of the failures of the communist system. Given that many of the most sustained and well argued criticisms of communism in the West came from neo-liberals, there was probably a natural inclination for many opposition intellectuals at least to take New Right ideas seriously.

In a country where intellectual values are respected and where the left had been thoroughly discredited it was therefore reasonable to expect that many would gravitate to a form of right wing thinking which went beyond the traditional elitist politics manifesting themselves elsewhere in eastern Europe.[6]

But if the Czech right had such a natural affinity with neo-liberalism and the parliamentary parties of the right were equally committed to it, why have calls to unite into one single party fallen on deaf ears? One possible answer is that though they all profess support for neo-liberalism, the largest, ODS, has failed to be neo-liberal in practice. In Chapter 5, where neo-liberalism is discussed in a different context, it is argued that Klaus shows signs of having only a primitive understanding of many important neo-liberal principles. It is also clearly true that ODS-led governments did not pursue such radical neo-liberal policies as rapid deregulation of the housing market, complete abolition of employment protection laws, drastic cuts in taxation and so on. But members of the other two parties formed part of these governments and could also see the logic of compromising aspects of neo-liberalism so as to avoid alienating public opinion. All ideology based parties in power fall prey to compromise, but this need not mean that their identity ceases to be defined in terms of the ideology from which deviations are made.

It may also be tempting to interpret the underlying reason for division in terms of corruption, and especially Klaus's seeming toleration of shady business practices arising out of the privatisation process. It is after all true that the coalition fell on a party financing scandal within ODS and calls by several senior members of the party for Klaus to take responsibility and resign.

But quite apart from ODA's own party financing scandals, ODA and US politicians occupied key posts in the government at the time when most of the corrupt practices are alleged to have been taking place.

Differences in this respect do not, in short, appear sufficiently substantial on their own to explain the continued divisions; although, as we will see, they may have caused serious friction in certain situations.

Division in the right appears to have much less to do with principle or standards of conduct than personality.

The general reasons for the emergence of personality based politics in a post-communist environment are well known. Loyalty to political parties only a few years old is difficult to secure. A strong and charismatic leader, and in his own way Klaus has certainly been that, is a bonus in all political environments. In the relatively immature post-communist Czech Republic the temptation for ODS to construct itself around a popular personality was too great to resist. This of course had one obvious corollary. If other political parties had problems with the ODS leader figure, the party which was built around him could not easily replace him with someone more suited to cross-party unity. Periodic calls throughout the 1990s for ODS to make itself more palatable to other parties by removing Klaus have thus missed the point. Voter loyalty to the ODS leader may well be greater than to ODS itself. Klaus was bigger than his own party.

This perhaps explains why Klaus has been such a powerful obstacle to unity in the political scene generally. There is one other, much less obvious factor, which has contributed to disunity on the right in particular.

The problem arises from the relationship between elite oriented policies and political parties aspiring to a broad voter base. Neo-liberal ideology may have acted as the glue holding together the politicians of the right at the elite level but neo-liberal policies are not easy to sell to the masses. One can perhaps popularise fiscal conservatism in terms of responsible government. But other policies have less immediate appeal. Imagine for example the difficulties of convincing a mass audience that the abolition of employment protection will increase the availability of jobs and, with the resultant productivity improvements, raise wages. On the popular level, such arguments are simply counter-intuitive and there are relatively few people in a population who will ever study them in sufficient detail to see their logic.[7]

The temptation for ODS to divert attention away from complex policy issues and onto the leader figure instead, could therefore have been a product of this problem.

Having been crowded out, as it were, of the mass market, ODA and US have conversely found a constituency among those elite sections of the public, especially those centred in Prague, which have been attracted to neo-liberal politics on intellectual grounds alone. Their voters did not need, and may even have resented, the presence of a strong personality as a means to securing support.

ODS on the one hand and ODA/US on the other, have attempted to sell a similar ideological package to an entirely different base of voters. This has reflected back on the make up of the parties themselves, giving them different political characters and specifically dividing them around the issue of Klaus's populistic, personality-based style of leadership.

The centre

The centre of the party political scene has been dominated by the Christian Democratic Union–Czechoslovak People's Party (KDU-ČSL). It has managed to sustain a loyal voter following from around 10 per cent of the electorate and until the ODS-ČSSD deal for mutual support in 1998 was the clear king-maker in Czech politics. It could give power to either the right or the left and always held open the threat to the rightist government that it did join that it would sink the coalition if its views were not respected.

It seems plausible to speculate that KDU owes some of its initial success to concerns among centre-left voters about the credibility of the Social Democrats who took a number of years to emerge as a strong force for political change. Voters who disliked both Klaus's personality and his right wing stance, but were unsure of the implications of voting for avowedly left wing parties, were naturally inclined to plump for the political centre. The party could also count on the support of those small but solid sections of the electorate that wanted to see a Christian slant to the political debate. It therefore had a distinctive constituency.

Like centrist Christian parties elsewhere it has been a force for moderation, suggesting that it supports some sort of social market economy and keeping alive, though not stressing, a religious influence on politics. Because of its previous association with right wing parties in government, it has sometimes been described as centre-right, but the temper of its parliamentary membership and some public statements to the contrary from senior KDU officials suggests that it does in fact remain firmly in the centre of Czech party politics.[8]

The left

The left of the political spectrum is made up of Prime Minister Miloš Zeman's Czech Social Democratic Party (ČSSD) and the almost unreconstructed[9] Communist Party of Bohemia and Moravia, the KSČM.

In contrast with most other such parties in the region, the KSČM has refused to shed the word communist from its title and has not remoulded itself as a labour or social democratic party of the European mainstream. It has none the less remained a solid force on the political scene, retaining the loyalty of around one in ten voters and even showing signs of picking up support at the end of the decade.

The reasons for this are to be found in the country's particular experience of communism and the kind of communist party which lost power in 1989.

Following the events of 1968, reformist figures, present or at least developing a presence in other ruling communist parties in the region, were expelled from the Czechoslovak communist party en masse.

Those few survivors who did harbour reformist tendencies kept quiet about them and probably felt less loyalty to the party than their hard-line counterparts. What was left in 1989, then, was a core of people whose loyalty was not just to the communist party but a communist party of a particular type. Reformers in other communist parties across the region could tell themselves and their sympathisers that times had changed, that it was now necessary to adjust left wing politics to a new environment. The Czech communist party was filled with people whose membership had been specifically defined in terms of opposition to reform. Many among the generation of post-1968 communists thus found themselves in a unique predicament. In other countries, communists could pacify their consciences, arguing that they had merely been constrained by the exigencies of the time. This may not have been entirely honest but it was at least possible as a modus vivendi. For Czech communists this avenue was basically closed.

There was also the fact of a strong tradition of mass support for communism dating back to the 1920s. This represents another crucial difference with the rest of the region. Tens of thousands of older voters in the 1990s had cast their ballots for communists in free elections in 1946. Some found it easy to admit that they had made a mistake. For others, the idea that they had been partly responsible for what had happened since the Second World War was probably too much to own up to. In addition, of course, although the Czech reform process was relatively soft in terms of its social impact, pensioners were particularly hard hit by the inflation of the early years.

For these reasons the communists have managed to sustain a solid, if ageing, voter support base in the absence of a fully fledged commitment to reform.

Since the KSČM has maintained an ambiguous stance on the communist past, cooperation with their much bigger brother on the Czech left, the Social Democrats, has proved problematic.

The ČSSD did not emerge as a powerful force until the 1996 elections, when it confounded opinion poll surveys and almost every political analyst in the country by taking 26 per cent of the vote. Polls had registered gains for the Social Democrats in the preceding months but this tally compared with just 7 per cent at the last elections four years earlier.

Since major social dislocation was not a distinctive feature of the early or middle parts of the Czech reform process – the biggest disruptions were in any case caused by the price deregulations which occurred in 1991, that is to say well before the 1992 elections – it seems reasonable to speculate that the belated rise of the Social Democrats was itself a product of the failure of the KSČM to emerge as a fully reformed party of the left. Democratic forces of the mainstream left, which could not throw in their lot with the communists, took time to regroup. Polish and Hungarian leftists could more quickly channel their sympathies into votes, since communist parties which had reconstructed themselves, complete with strong organisational structures, simply appeared more palatable. The Czech Social Democrats had to build themselves up from scratch and, in contrast with other parties which had to do the same, faced the confusing presence of an established rival with which they would have to compete for votes.[10]

ČSSD, a fully integrated party of the democratic left drawing on strong social democratic traditions in Czech history, therefore has had to define itself in terms of its opposition to the neo-liberalism of the right but also in even clearer distinction to the KSČM. As a matter of party policy it has eschewed all formal, governmental-level, contact with the communists and presents itself as much as a product of the post-communist era as its opponents on the right. It has been remarkably successful in achieving this aim.

The last piece of the party political puzzle is made up of the far-right Republicans. The party is routinely referred to as far right but in reality belies definition along the standard right–left demarcation line.[11]

The Republicans exemplify those small but persistent forces in Czech society which militate against the development of liberal democracy and which seek to foster attitudes and dispositions inappropriate for its successful establishment.

The party is hostile to Gypsies, Germans and foreign influence in general and it has relied both on public support for such sentiments and the maverick behaviour of its leader, Miroslav Sládek, rather than substantial policy initiatives in order to sustain support.

Like the communists, no mainstream party has officially cooperated with the Republicans so far, and the happiest, though perhaps short-lived, consequence of the on-going political crisis in the country is that the party failed to cross the 5 per cent threshold necessary for parliamentary representation at the 1998 elections.

We can see from this survey how the alignment of forces in Czech party politics has been forged at a crucial level on terms defined by the transition from communism.

It should also be clear why consolidation has been so problematic and how fragile were the bonds holding together the three-party centre-right coalition which was in power for the substantial part of the period after independence.

Against this background the narrative of events beginning with the end of Klaus's premiership and ending with the ČSSD-ODS power sharing agreement begins to make more sense.

The fall of the Klaus government

It required only a couple of sharp blows to bring the ODS-ODA-KDU-ČSL government to an end in an orgy of public recrimination. Both landed heavily on Klaus and wrought nearly fatal damage to his political career.

The first and most important came with the downturn in the economy. If there was one reason why ODA and KDU-ČSL were prepared to put aside their long standing distaste for Klaus's style of leadership (apart from the fact that in so doing they were ministers in government rather than deputies in opposition), it was the presumed success of the economy under his stewardship. It felt good to be part of a government which appeared to have conducted a transformation process without causing major social distress. Western governments and investors lavished praise on the Czech government. They could all take credit for what was routinely described as the Czech economic miracle.

There had of course been some voices in the background suggesting that low unemployment existed because of weak restructuring; that the trade deficit was evidence that Czech companies were increasingly unable to compete; that only a corrupt or lucky few had seen real benefits from voucher privatisation; that the capital markets were riddled with insider trading; that the banks were waist deep in bad debt.

And yet solid growth and low unemployment figures continued to roll in. It seemed that Klaus, who had built his political career around an image of superior economic know-how, really was possessed of a Midas touch, turning a rusty old communist economy into gold with breathtaking speed and efficiency.

Two economic reform packages and the collapse of the Czech crown (currency) in April and May 1997 put paid to all that. This was not merely a harbinger of problems to come. Klaus and his government had been subjected to public humiliation at home and abroad.[12]

This can only have come as a massive blow, reverberating back on the confidence of the coalition and causing its members to reevaluate their positions. As soon as Klaus's credibility as a competent manager of economic reform was called into question, so was the overriding reason why ODA, KDU and, as it turned out, senior members of his own party, had been prepared to put up with him.

From this point on, the glue which had held these awkward partners together began to melt. It was only a matter of time before something would pull them irreconcilably apart, and no issue presented a greater opportunity than a party financing scandal.

On the face of it, the accusations, some groundless others substantiated, that ODS had been using corrupt means to secure party funds does not appear as the kind of issue on which to break up a government. ODA, for one, had some questionable dealings of its own to answer for. And party financing scandals are not exactly rare in the West either.

But it is important to see that corruption generally and Klaus's arrogant dismissal of its importance in particular was rapidly becoming the political issue of the day. No criticism was more forthcoming from the mouths of opposition politicians or the pens of newspaper columnists than that Klaus was insufferably conceited as an individual and that his reform project had handed economic power to shady investment funds if not outright crooks. While it was a substantive issue in its own right, the issue of corruption was also a particularly potent weapon against Klaus personally, touching on his moral probity as well as his competence.

As it turned out, the demise of the coalition was not initiated by disgruntlement from ODA or even KDU, it came from inside ODS itself. The dominance of Klaus's personality and all that that entailed, was not, it seemed, merely a wedge driving apart the parties of the right, it acted as a destabilising factor for other personalities within his ODS itself. If ODS truly was built around one individual then the ambitions of others were going to be stymied by his presence. Taking a less cynical view, if committed neo-liberals within ODS felt that their brand of politics was being discredited by a leader prepared to put up with or even be complicit in corruption, there would come a point at which a challenge would have to be mounted.

Either way, the challenge failed to dislodge Klaus as party leader, but it did bring the government down and redraw the battle lines in Czech politics. Before we proceed to discuss what came next, and in particular the implications of the extraordinary deal struck between the two titanic enemy figures of ODS and ČSSD in the middle of the following year, we need to see how the troublesome developments in the party system had a profound effect on the presidency, the other main player in Czech politics since the revolution.

The President joins the fray

The Czech President is formally elected by parliament and his role is mainly ceremonial. He receives foreign visitors, heads the armed forces, names the prime minister, and in certain circumstances he can dissolve the lower house. He must sign bills into law and can return those he does not like to parliament. However, this power can be overridden by a simple majority and he cannot call referenda or rule by decree – a deficiency which some argued had serious consequences in the period leading up to the end of Czechoslovakia.

There is nothing so different here from the situation prevailing in several western countries, and the role of the president in the Czech political system after the Velvet Revolution would be unexceptional but for the fact that the office is occupied by Václav Havel.

The effectiveness of a symbolic presidency is bound to be strongly determined by the charisma and moral authority of the person who holds it. In the Czechoslovak and subsequently Czech case, the incumbent has for most of the time towered above all others on the political scene for the respect he has been able to command among his people.

Although Havel was initially not widely known outside dissident and intellectual circles at home – except as an enemy of the people – he quickly came to personify the revolution against communism. In the absence of any party loyalties outside rump supporters of the communists, personalities, as we have seen, were always likely to assume great importance in the immediate aftermath of the revolution. Havel's informal style, his casual dress sense, his public association with foreign rock stars and the obvious respect he enjoyed among western dignitaries marked a huge break with the grey suited leaders of the past. He was serious but unpretentious and when the truth about his uncompromising and brave stand against communism became more widely known he acquired an iconic if not actually heroic status among the public. This was the foundation of his moral authority and until 1998, at least, it has been durable. While many Czechs react badly to what they sometimes see as moral lecturing from other dissidents, Havel's role as guardian of the national conscience has been largely unchallenged. Few party politicians have dared to question this authority, and in so far as they have this has happened in the latter part of the decade and for reasons which we will shortly discuss.

With this authority intact, the main role that Havel could play in the early years after the revolution was to offer his personal stamp of approval to legitimise the new institutions of the emerging democratic society. Havel could hold the line until loyalties to parties and democratic processes were better defined and more clearly understood.[13]

The difficulty he faced was bridging the transition between those early years when there was broad acceptance by himself, the public and the emerging political parties that the authority of the presidency needed to be substantial, and later years when other political actors felt sufficiently established not to need, and even to resent, the crutch Havel felt obliged to provide.

The challenge can be summarised thus: Havel established himself as a non-party president at a time when there were no substantial parties. As that situation changed so would the scope of the presidency. And yet on top of this, Havel had tagged on the idea of a presidency conceived as the moral guardian of the nation. He did not see this as a purely temporary role. This would complicate Havel's efforts to extricate himself entirely from the party political fray, as the events of the end of 1997 showed up all too starkly.

The high point of presidential engagement in domestic politics following the split with Slovakia came at the very time the coalition fell apart. Following on from what has just been said this was perfectly predictable. While the party political system was still holding together Havel would have been exceeding the limits of a non-partisan presidency had he waged public confrontation with any of the mainstream politicians. He had always been ready to express his misgivings about the republicans and the communists, but neither was in government nor in substantial opposition and both were conceived by the main political parties as outcasts.

But his role in appointing the non-party government led by (an obviously reluctant) central bank governor Josef Tošovský is not the most significant event in the development of the presidency. It was Havel's job in that kind of situation to broker some sort of deal and everyone, Klaus included, accepted that. In any case, even though the Tošovský government was opposed by Klaus, no one had any acceptable alternatives. Of more lasting significance was the way the whole affair brought Havel's visceral dislike for Klaus and his style of politics into the open.

Even before the resignation of the government, President Havel's participation in a campaign to dump Klaus was being mooted by ODS officials as senior as the party's deputy chairman Miroslav Macek.

> I believe that most KDU-ČSL members are convinced that if the current coalition ruled with a different premier they would be able, on the one hand, to better implement their socially-orientated market economy, or at least something they see under this title, while at the same time the coalition would have wider support... [this is] also true, and maybe more so, for the Castle.[14]

By the end of the month the government was finished. At a meeting of the coalition leaders at Lány on November 30, Havel announced that the government had resigned and that talks, which he would lead, on the formation of a new one would be postponed until the middle of December.

At the Lány announcement, Havel's treatment of Klaus was peremptory to the point of rudeness. He simply said 'I would like to thank Václav Klaus for all he has done for this country,'[15] and nothing else. This was damning Klaus with faint praise indeed. Just over a week later in a speech on December 9 to both houses of parliament and diplomats (see Chapter 3) he tore into the record of the outgoing government. He did so moreover in explicitly moral terms. Klaus was not mentioned by name but it was clear, especially to Klaus himself, who issued a stinging rebuke to the president the same day, that he was the object of the attack.

Once Havel had entered the political fray so clearly he would never quite manage to pull himself out. In making his opposition to Klaus as crystal clear as he did he could no longer be seen, especially by Klaus's many supporters, as a non-partisan president. He had exhausted the limits of the type of presidency he wanted to establish. The affair highlights the problem of a presidential system built around moral authority rather than one purely construed in terms of guardianship of the state and the performance of associated duties. Since Havel's modus vivendi was strong moral leadership he could not forever refrain from opposition to a style of party political government which he rejected on profoundly moral grounds. But once he had plunged into the political debate he had taken what he had always seen as a non-partisan presidency with him. The contradiction was palpable.

The moral authority that the Czech president has been able to wield in the years after the Velvet Revolution is unique in the region. Though his stature as a unifying figure for Czech, though most definitely not for Czechoslovak, society may well have been advantageous, it was clearly a double edged sword in the transitional period. The very blurring of Havel's constitutional powers with his personal authority has left the Czech Republic with an important unanswered question in so far as the viability of post-communist constitutional arrangements are concerned. All other major offices in the high echelons of political power have been subject to change. The country is aware of what it feels like to have a different prime minister, a different government, a new parliament. But it is not yet clear what the country would be like with a president whose name is not Václav Havel. The very continuity that Havel has brought to the constitutional apparatus is a potential, though in reality probably small, destabilising factor.

The next Czech president will have to try and make sense of this without possessing Havel's unique position in Czech society. Until this problem has been fully resolved it will not be possible to close the book on the Czech Republic's constitutional transition from communism.

The 1998 elections – the new deal

If the fall of the Klaus government and the appointment of Josef Tošovský's caretaker administration showed up the fragility of post-communist party politics and the tenuous nature of Havel's presidency, those who hoped for a fresh start or a resolution at the 1998 elections were in for a shock.

One right wing party, ODA, virtually disappeared from sight, and its position was taken at the elections by the Freedom Union (US) which was led by the anti-Klaus faction in ODS. US took 8.6 per cent of the vote, with ODS on 27.7. Although the name of the smaller party of the right had changed, the pattern of one large and one small New Right party was sustained. Despite heavy losses in the opinion polls earlier in the year ODS had retrieved almost all of its support, adding further weight to the assessment that the party's pool of voters remained fiercely loyal to the personality of Václav Klaus.[16]

Interestingly, with KDU-ČSL taking 9.0 per cent of the vote and 20 seats, the combined forces of the centre and right could have formed a small majority government. But differences emanating from the fall of the government in 1997 precluded any such reconciliation.

The Social Democrats were the clear winners, taking just over 32 per cent, and the communists edged up a few notches on the 1996 elections with 11.03 per cent.

The election was bad news for two of the maverick parties, which perhaps explains the rise in support for the communists. The Republicans, whose leader had been in jail until cleared of inciting racial hatred, failed to cross the 5 per cent threshold, as did the pensioners party which had shown signs earlier in the year of mounting a serious challenge.[17]

The make-up of the party political scene described above now showed itself for all to see.

The right was even less able to consolidate because US had been founded ultra-specifically in opposition to Klaus's style of leadership. At the same time, ODS was more clearly built around Klaus than at any time during the decade, making any suggestion of his being removed from within utterly implausible. The left was still characterised by a stronger, even enhanced social democratic party which could not, it seemed, govern alone, and even if it had taken the unlikely step of joining with the communists would still have been unable to form a majority.

The centre, which on paper and in the absence of all the other causes of division, could have provided a majority to either the right or the left, refused to do so. It did not want to work with Klaus, whose personal style it rejected as much as anyone else, and as a Christian centrist party refused point blank to consider a centre left government with communist participation.

In terms of forming a majority government, the options appeared to be limited. The only other vaguely realistic prospect was for a government between KDU, US and the Social Democrats. But US, probably the most right wing party on the scene, would have risked a lot by throwing its weight behind ČSSD. It had just been formed as an avowedly rightist party and any deal with the left could have resulted in the evaporation of its support and the potential revival of ODA or yet another small right wing party to cater for its undoubtedly disillusioned constituency.

To the surprise of almost everyone and the consternation of the smaller parties, the 'resolution' to the problem was a formal pact signed between the two largest parties whose practical result was that Klaus's ODS agreed to tolerate a minority Social Democratic government.[18]

The July 9 agreement allowed whichever of the two parties had won an election to govern without fear of losing a vote of confidence. The party which came second would have the right to chair both chambers of the parliament and to head up certain parliamentary committees. It also specifically addressed the matter of reforming the constitution to provide for more stable outcomes at parliamentary elections.

The agreement provoked fury from KDU-ČSL and US who, wrongly, called it unconstitutional and worryingly called it undemocratic. This latter accusation was the more serious. For the first time since 1989, the democratic credentials of a Czech government had been called into question by mainstream parties.

Although everyone could see that there was no easy solution, Havel spoke for many in offering his very grudging acceptance:

I feel it my obligation now to express my fears over this very unusual solution, which is an agreement between a left-wing party that has for years fought against a government of alleged right-wing embezzlers and a right-wing party which has called for mobilisation against the left as allegedly attempting a return to socialism.[19]

The main problem, as Havel hinted, was that it called into question the integrity of the two biggest Czech parties and also the value of the franchise itself. Politicians are frequently accused of breaking election promises or reneging on their principles to keep power, but there can be few clearer examples of naked personal ambition dictating the fortunes of a country's political development than this.

In view of the newness of the ties between voter and politician and the relatively short Czech experience of parliamentary democracy, the dangers of the agreement were obvious. If the biggest right wing and left wing parties had agreed to squeeze out the smaller parties and ensure that whichever of them was the bigger would govern with the other's agreement, all the while remaining mindful of the opposition's ideological sensitivities, what was the real point of going to vote at all?[20]

Nevertheless, there were also dangers in maintaining an arrangement which did not seem to be able to provide the country with stable government. Havel spoke of the intention to change the constitution as being conducted 'under the pretence' of being interested in political stability.

From the perspective of the personalities involved he was undoubtedly right, but from an institutional viewpoint it also seems possible to interpret the affair as a nascent democracy solving its own internal problems.

Perhaps the most honest solution would have been to begin immediate drafting of new electoral rules, put them to a vote in parliament and then a referendum.[21] This could have taken some sting out of the charges of cynicism but it obviously would not have solved the immediate problem. The country needed a government.

Although it is unlikely to happen, the fastest way to move the country on and improve the chances of stability in Czech politics would clearly be the retirement of Václav Klaus. His continued presence militates against the unity of the right and the formation of government with the centrists. The more probable alternative of changing the electoral system with a stronger bias against small parties may improve matters. But it would be naive to say this with any great certainty.

There are strong underlying reasons for the existence of the parties in their current form. Legislating against the smaller of them may provoke either a backlash, which improves their support and actually further fragments the political scene, or it may provoke bitterness and disillusion. At a time of rising unemployment and economic stagnation the risks may be even greater. It is of course precisely against this sort of economic background that extremist parties come into their own. The consequences are not clear. On top of all this, the most visible political actor, President Havel, represents a force for continuity and simultaneously the threat by his potential demise of a degree of uncertainty.

Ten years after the Velvet Revolution, many important questions about the direction of Czech politics remain unanswered. Some of the most important elements of the system – the parties and the presidency – are clearly framed in terms of the legacy of a particular communist past. The consequent difficulties which this has led to in governing the Czech Republic have raised questions about the way in which the democratic system is organised. But let us retain a sense of proportion here. The principle that whatever changes do occur should be carried through by popular consent is firmly established. The Masarykian legacy alluded to at the beginning has not disappeared. Democracy is not under threat in the Czech Republic. The transition, nevertheless, goes on.

5
The Economy – Capitalism without Property

By the middle of the 1990s a new phrase was creeping into the lexicon of Václav Klaus's speeches. True to form, it had a confident and reassuring ring to it. The Czech Republic had entered 'The post-transformation phase.' Switching into metaphor, eastern Europe's most market friendly, post-communist reformer explained the basic idea thus: 'Using my standard analogy and describing the three consecutive transformation stages as waiting in a hospital ward, undergoing surgery and recovering in a rehabilitation center, I can assure you that we have made it to the rehabilitation center. And we are in pretty good shape now.'[1] The speech was made in November 1994. Having suffered a severe and traumatic relapse the patient is recovering in intensive care. The head doctor has been demoted and he has struck up a curious association with a group whose practices he once dismissed as old-fashioned.

By the end of the decade the Czech economy was back in deep recession. While Hungary and Poland were posting growth rates of between 3 and 6 per cent a piece, the one time flagship of east European reform was leaking badly. Major banks tottered on the brink of collapse, unemployment was rising fast, industrial production was plummeting and for much of the second half of the decade the PX50 stock index had stood at less than half the value of when it was set up.[2] Far from catching up with the West, the Czech Republic was slipping further behind.

International organisations were queuing up to change their minds about the Czech economic miracle.[3] It was even being suggested that the Czech economy had failed to recover from the downturn suffered

Table 5.1 Development of contentment with the economic transformation process, 1993–8 (%)

	93/03	94/03	95/02	96/03	96/10	97/02	97/03	97/05	97/07	98/02
Successful	26	28	31	28	23	18	20	10	8	5
Neither	42	47	44	43	42	44	34	40	32	31
Unsuccessful	26	17	18	19	26	27	36	44	53	61
Don't know	6	8	7	10	9	11	10	6	7	3

Margin of error +/– 3 per cent
Source: IVVM

by all transitional countries in the early 1990s, putting the country's estimated gross domestic product by the end of 1998 at below the level reached at the fall of communism. Poland was up around 17 per cent, Slovakia had reached parity and among comparable countries only Hungary had fared worse.

As we will see, the reliability of such figures is questionable, but they could only contribute to a decidedly gloomy mood among a people who had become used to praise from abroad for their economic achievements.

Opinion poll evidence paints a sombre picture. (See Table 5.1.)

From a high in 1995 of 31 per cent among those who thought the reform process had gone well, the number of positive respondents had fallen to a mere 5 per cent, 1 in 20, by 1998. More or less the reverse trend was apparent among the pessimists, whose numbers had risen from a low of 17 per cent in 1994 to a massive 61 per cent at the end of the period measured.

It would be foolish to place all the blame at the door of one individual but there can be no doubt that investigations into what went wrong should begin with the policies and ideological convictions of the undisputed master of the Czech road to economic reform, Václav Klaus.

As Czechoslovak finance minister, Czech Prime Minister and, after the break-up of Czechoslovakia on January 1, 1993, Prime Minister of the Czech Republic, Václav Klaus has dominated the thinking and practice of economic policy since the Velvet Revolution. His free market, small government, anti-state rhetoric earned him a reputation in the West as the most committed and able reformer in the former Soviet bloc. Margaret Thatcher, among many words of praise, once referred to him as her 'other favourite prime minister'.[4] He counted

among his friends free market Nobel laureates, and the more syco-
phantic of his supporters (and there were many sycophants) were sure
it was only a matter of time before his distinctive contribution to
economic thinking would be rewarded with a Nobel prize of his own.

In relating the main events of the economic reform process at the
end of the decade it is easy to lose sight of just how self-assured Klaus
and his coterie actually were. The following example (with my italics)
is pure Klaus.

> At the individual level, wages (and other forms of income) should
> temporarily lag behind productivity; at the national level, the
> exchange rate (dramatically lowered before foreign trade liberal-
> isation) must temporarily stay below purchasing power parity.
> Those two principles are the basis of *my recently formulated 'hypoth-
> esis of two transformation cushions'*.[5]

Added to Klaus's practice of quoting himself in his own speeches,
this sort of argument elevated to the level of 'hypothesis' and 'theory'
(to make it sound grand, presumably) paints a clear picture of a man
convinced of his own status as an economic guru.[6] (In any case, wages
actually far outpaced productivity for many of the early years and the
second transformation cushion, the exchange rate, was partly eroded
in the form of real appreciation as Klaus became obsessed with the
currency as an anchor for the economy.)

Klaus is an easy target but criticism is invited by his sheer arrogance.
While acknowledging that what was being attempted in eastern Europe had never been tried before – turning an aquarium into fish soup
was far easier than achieving the reverse – Klaus and those around him
closed their minds to criticism of any kind. Journalists, even those
generally sympathetic to the free market reform process, were dismissed as the 'worst enemies of mankind', and one of the most grat-
ingly common responses to pertinent questions at government news
conferences, was for Klaus to simply tell the offending reporter, 'that
isn't the question'.[7] This is of course a subsidiary point, but it seems
clear that Klaus lacked the appropriate temperament for a reformer in
his position. Resolution and firm leadership in the face of public
opinion were obviously necessary to create the swell of enthusiasm
required to mobilise a society used to passivity. But so was an ability to
fine tune, occasionally rethink and go back to the drawing board. If, by

some miracle, Klaus had got everything right from the start, the consequences may not have been so dire. Unfortunately, his comprehension of some crucial elements of the capitalist system was too primitive for him to manage the job on his own.

In trying to explain how it all went so badly wrong some arguments focus on the structural dynamics of political change; Klaus found himself answering to and compromising with industrial and electoral interest groups. He had to soften his hardline theory to sustain the general thrust of his revolution. While other reform minded governments in eastern Europe lost out in an early backlash against the social costs of 'shock therapy', Klaus played an altogether different game, never pushing reform too sharply against the grain of popular feeling. He compromised the reform process but in so doing avoided handing the initiative to retrograde forces that would have abandoned free market policies altogether. This version of events, which we will explore in more detail, can be given a generous or a cynical slant. On the positive side, Klaus can be pitted as a responsible reformer. No revolution can ever live up to all of its ideals, if only because circumstances always change in ways that can never be seen in advance. Better to accept unwelcome compromise and move at a surer pace than risk the catastrophe of total derailment.

A less flattering interpretation paints a picture of Klaus as the consummate politician. The extreme version portrays him as a man whose alliance with the cause of neo-liberalism was little more than an appropriation of the language of the time. The late 1980s and early 1990s represented the high point of New Right dominance. Ready-made free market arguments were there for the picking. Transplanting them onto the impoverished body politic of post-communist society was always going to be an easy job. Foreign governments, the ones that mattered at least, would rush to congratulate him and opponents at home could be easily dismissed as half hearted. Neo-liberal ideology made Klaus distinctive, setting him apart from the compromisers and in direct opposition to what had gone before; the perfect vehicle for political advancement.

There may be much of value in this line of thinking, and no explanation of Klaus the politician would be complete without it. But the core failing in his economic transformation experiment is to be found at the conceptual level, as we can see in discussion of the distinctively Czech contribution to east European transition: voucher or coupon, privatisation.

Czech privatisation – justice with efficiency

In Klaus's own words: 'We tried to look for privatisation techniques that would be faster than the standard ones. For that reason we prepared and successfully implemented a non-standard technique called voucher privatisation. The idea is relatively simple. It is based on selling vouchers (quasi-money usable only in the privatisation process) to all citizens of the country at a symbolic price. The citizens subsequently use their vouchers to buy shares of privatised firms. [Privatisation by voucher turned more than 75 per cent of Czech adults into shareholders. Each of them now owns shares in either some of the 1,500 privatised companies or in some of the investment privatisation funds]...Our nonstandard voucher privatisation proved to be rapid and efficient.'[8]

Voucher books were sold at an affordable price to all citizens over the age of 18. Bids would be put in for companies. If demand outstripped supply, a new round would take place for higher stakes. Eventually, a 'price' would be found and vouchers would be exchanged for shares. The more the demand, the greater the price in terms of vouchers used up.

The voucher method had a number of seemingly obvious advantages. The most politically significant was that it promised to avoid handing the country's industry over to those who had profited under communism – apparatchiks and blackmarketeers. In fact it turned out to be a perfect way of keeping it in the clutches of the nomenklatura because enterprise managers were charged with the task of drawing up privatisation projects themselves. Clearly they were not going to recommend solutions which did not include them as crucial constituent elements. Also, many of those who could not stay in politics after 1989 were moved into company management positions. This was simply the easiest, most painless method of clearing the way for the new politicians and state officials.[9] Nevertheless, the theory was sound and although coupon privatisation did result in many members of the old elite getting their hands on the new property, it need not have done if certain practices had been different.

Another advantage at the popular level was that foreigners were prevented from cherry picking. The revolution was just and it delivered its fruits directly into the arms of the people. In common with the state industry and council house privatisations of Thatcher's Britain in the 1980s, citizens were given a personal stake in the reform process. They were simultaneously given a tangible reason to be grateful to Klaus. It was also, as Klaus says, incredibly fast. (The second wave in the two wave voucher privatisation process was held up by the split with Slovakia, but this cannot be seen as something essential to voucher privatisation itself.) Whereas direct sales would have been conducted in dozens of separate auctions, huge sections of the economy were sold off in just two waves.[10]

The faster the economy could be turned over to private hands, and by 1994 the EBRD was saying that the private sector already accounted for 65 per cent of GDP, the sooner the country would see the rewards from efficiency savings and entrpreneurship. Even if foreigners were excluded in the first stage, it was a good signal that the government meant business and it made the Prague Stock Exchange among the most heavily capitalised in the region.

Most of the drawbacks were less immediately apparent. The biggest early problem was getting a people unused to capitalism interested in obtaining shares at all. The first few months of the first wave were decidedly sluggish. Only a few hundred thousand people bought voucher books, threatening to sink Klaus's experiment before it had even set sail. But the market, in the form of a 28-year-old called Viktor Kožený, provided its own solution.

Recognising that a money for nothing guarantee was the only sure way to galvanise the people into action, Kožený offered anyone who bought into his scheme a 10 for 1 return within a year and a day. As other investment funds sprang up to imitate Kožený's example, the voucher books poured forth in their millions. The Klausian revolution was safe, or so it seemed.

Before looking at the consequences of the dominance of investment funds in the coupon privatisation process, a few other problems are worth noting at this point. One was that the coupon method generated no money for the companies. Badly needed capital for investment did not flow in from this form of privatisation. Neither did it create much money for the government, a luxury that debt ridden reforming countries such as Poland and Hungary could not afford.

A more urgent problem was that excluding foreigners obviously blocked the inflow of foreign direct investment. Also, giving everyone the equal right to own shares granted those with a strong commitment to, or a sound understanding of, a particular business no advantage over those whose choice was based on a whim. Not only was ownership incredibly diverse – which made corporate governance problematic to say the least – there was no certainty that owners knew or cared about the companies they had partially acquired. Klaus was unimpressed by such concerns.

It is no exaggeration to say that we consider it unnecessary to design techniques and legislation with the objective of selecting perfect owners. An objective like that is far beyond the capacity of post-communist, or any other, governments, and the initial owners may not be the final ones, anyway.[11]

Unpacking the assumptions contained in this thought brings us to the heart of the Klausian experiment. It shows us why he was given so much time by western governments, how he succeeded in persuading his domestic audience of his truly radical vision, and in what it lacks, it shows us where he went wrong.

Klaus's thinking was based on a sound enough premise: the all important company restructuring which would eventually bring the real fruits of a free enterprise economy depended not on who initially owned this or that company, their competence or suitability to do the job.[12] What mattered was that the correct general conditions should be created in the economy as a whole. Competition would force companies into a narrow range of options regardless of who owned them. Those who tried to buck the market would suffer the consequences of its unyielding discipline.

In the best cases, competent ownership would filter through from the capital markets where those with the wherewithal and enthusiasm to take over companies they knew they could improve would buy out the original owners. In the worst cases bad owners would suffer the 'freedom of exit' from economic life known outside the economic faculties as bankruptcy. The market economy would do efficiently what the blunt instrument of government intervention had never been able to manage. The argument came directly out of the texts of Friedman and Becker and was unimpeachable as far as it went. The problem was where it did not go.

Klaus and his critics focused heavily on questions of *who* would end up owning state property – Klaus to say that it didn't initially matter and his critics, most commonly, to point up the dominance on company boards of former communists on the one hand and banks and investment funds on the other. Far less attention was devoted to what ownership actually meant and which conditions were necessary to make markets work.

Market discipline requires a functioning market and vice versa. In Klaus's admirable attempt to simply get the economy into private hands, he staked all on the creation of an environment that would provide its own solutions to corporate governance, suitable ownership and, through this, company restructuring. Everything that was essential to the Klausian reform process depended on his willingness and ability to accomplish this task. He failed and there is compelling evidence that the failure can be traced back to a conceptual flaw in his thinking.

As Klaus never failed to remind his various audiences, what was being attempted in eastern Europe had no historical parallel. The most inspiring reference point for mass privatisation was of course the Thatcher period in Britain in the 1980s. But Klaus was well aware, at least he often said it, of the fundamental difference between privatising industries into an economy already furnished with the traditions and infrastructure of a developed market economy and the task facing eastern Europe. British people knew what private property meant. They knew how to go to court to protect their property and there were well defined laws and strong enforcement agencies to uphold their claims.

> Privatization in our case does not mean the standard shift of property rights between two (or more) well-defined economic agents but the establishment of a property rights structure that was previously either non-existent or very strange. Privatizing in the West may be viewed as a 'reform' process. In the East, however, privatization is the most fundamental objective of a systemic transformation.[13]

The essence of the problem was that neither in the regulation of the capital markets, the creation of well funded civil law enforcement agencies to guarantee contracts, or the promulgation of an efficient bankruptcy law did Klaus and his governments create the conditions in which the famous discipline of the market-place could function. It is important to see that this in fact represents the root of the problem with the Klausian experiment.

At the most fundamental level there is evidence that Klaus frequently paid lip service to the importance of property rights but rarely went further. At the very least it is clear that the issue did not figure very highly in his list of priorities. Before bringing forward illustrations to support this point let us pause to reflect further on this.

What is private property? A simple but misleading and incomplete answer is to say that it is anything that is not directly or indirectly held by the state. If the state sells a shop to its manager, something has passed from state hands into private hands. This is the stage of private possession but it is not yet private property. Suppose the manager's deputy comes into the shop while the new 'owner' is on holiday. He turfs out the staff, installs his own people and changes the name on the door to his own. On his return, the original 'owner' bristles with indignation and rushes to produce his buy agreement with the state. But his erstwhile deputy says it is a forgery. If it comes down to pieces of paper he has his own 'proof' that he is the real owner. The police are called but they are no better equipped to solve the problem, not having easy access to the official records and in any case too busy with muggings and burglaries to bother. They suggest that the problem be solved in the civil courts. Although the plaintiff has never heard of civil courts he catches on fast and sets his lawyers onto the case. In the meantime, the shop's inventory starts depleting rapidly; bank reserves go the same way. The day before the judgment is passed down, the usurper, who is found guilty and ordered to leave the premises and pay a big fine, departs with a sack of money to set up new premises on an island in the sun. Finally, private property rights have been established but the delay has bankrupted the owner.

The point should be clear. Private property is not something that simply arises when the state dumps its holdings onto an unsuspecting public. Neither does it emerge spontaneously.[14] (The spontaneous answer is the rule of the mob. Ownership is 'proved' by the size of the stick one is able to wield.) Private property is inextricably bound up with the state. In the absence of crystal clear, fast procedures for establishing who owns what and where their rights begin and end, there may be *possession* but there is not private property. The state therefore does not simply *guarantee* private property by establishing laws and enforcement mechanisms, in so doing it *creates* it.[15]

On some occasions, Klaus has come close to recognising what is at stake here. The following extract is a real rarity and forms only a fleeting part of the speech from which it is taken.

> I am a true believer in the Hayekian idea of evolutionary formation of all complex human institutions, but I know that after the institutions of one social system are dismantled the institutional vacuum must be rapidly filled with new rules of the game, both formal (laws) and informal (patterns of conduct); otherwise chaos and anarchy start to govern society.[16]

If only this line of thinking had been followed through.

Columnist Jan Macháček, amongst others, noted a clear tendency for Klaus to surround himself with economists, showing 'an ostentatious disdain for lawyers'.[17] In an article expounding on Klaus's populism, Macháček continues: 'Klaus's proclaimed liberalism reflected in practice in mainly the fight against interest groups or professional lobbies which, moreover, was verbal more than practical. There existed at least one interest group which was supported by Klaus – the lobby of opponents of the regulation of capital markets. Klaus's contemptuous approach to the protection of minor shareholders was an unprecedently serious mistake. This not only led to clear cases of fraud and thefts but to a situation in which hundreds of thousands of people were robbed of property in coupon privatisation...'[18]

The extraordinary conclusion that Macháček and many others were drawn to was that Klaus had made close to the whole adult population into small shareholders but had failed to provide the legal setting which gave that property ownership meaning. People had possession of shares but they did not have real ownership. When bigger fish smelt blood they could and frequently did snap up their prey with virtual impunity.

A new word, 'tunnelling', became part of the household vocabulary. It described the practice, often by politically well connected, shady investment groups, of hollowing out companies, transferring assets into cash, and sometimes simply siphoning money abroad. The cynical explanation as to how this sort of chicanery was allowed to go on is hinted at in Macháček's analysis above. Klaus became beholden to a new class of investors – itself the product of the dominance of investment funds in the voucher privatisation system – whose *raison d'être* was the power they had as little known insiders in a Wild West landscape ruled by the strong. The introduction of transparency, rule enforcement mechanisms and clearly defined rights for ordinary shareholders would have dissolved the basis for their existence and with it important clients of the ruling party.

It is hard to be sure precisely what made Klaus so hostile to a rule based market environment. But, as has been hinted above, it is not necessary to impute to Klaus such base motives.

Tomáš Ježek, then the Chairman of the Prague Stock Exchange, was in no doubt what was at work here. Asked in 1997 why a stock exchange watchdog had not been established years earlier he said: 'Ask the government. Why didn't Klaus want it. Because of this utterly false idea of liberalism/laissez-faire.'[19] In the same year, Klaus gave his own view on the matter, simply telling an economic forum in the United States: 'I am not very optimistic about regulation in general and about regulation of capital markets in particular.'[20]

This was a theme repeated in one form or another over the course of the whole decade. Klaus, who for most of his life had been accustomed to arguing his case against communists and social democrats, understood the free market argument well but only in its broadest formulation. The debate he had engaged in during the 1980s was with people who had no understanding of even the most general benefits of a market economy. In his readings of Friedman and Becker and in his appreciation of the reforms of Margaret Thatcher in Britain he showed a strong understanding of the dangers of creeping socialism via over-regulation in the West. His writings show a consistent and impeccably New Right distrust of the role of the state in building a successful economy. But he failed to appreciate the all important subtleties. When his mentors in the West spoke out against regulation they were lamenting the straitjacket it was imposing on free individuals to dispose of their property as they saw fit. By Laissez-faire they meant laissez-faire once private property had been firmly established as an inalienable right. Klaus rejected the word 'regulation' with no appreciation that in the context of shareholders, the capital markets and privatisation generally, regulation meant nothing more or less than the true establishment of private property. The rights he left his nation of so-called property holders with were anything but inalienable.

Lacking the secure base of firmly established property rights, the market conditions which Klaus's rapid privatisation programme depended upon simply did not arise. An investor looking at the prospect of buying shares had to reckon with the fact that he did not always know who the other owners were or that he could not get reliable information on the company's financial health. If he took a minority stake, what was to stop other owners getting together, sending the company's funds into their own bank accounts and rendering his shares worthless?

The voucher programme put vast tracts of the economy onto a stock market which individuals and investors were afraid to trade on. New and better owners were as likely to get control of Czech companies as were charlatans or insiders often with links to the old regime. Klaus had rightly said that the government could not pick perfect owners at the outset of the privatisation programme but he prevented the emergence of a viable capital market which would do the job instead. The result was that company restructuring was dictated neither by government nor by markets. The gap was filled by an awful mishmash of something in between. The consequences for the economy were obvious.

This picture of failure at the most basic level is brought into sharper focus in consideration of the role of the investment funds which, as we have seen, saved the coupon privatisation process from early disaster.

The first major problem with the privatisation investment funds (IPFs) was a contradiction between the roles they were expected to play by the lawyers who designed their rules of conduct and the economists who had a conception of their role in the wider economy.

The lawyers preparing the legislation for IPFs and investment companies were inspired by EU and US legislation and understood them as tools of collective investment. IPFs were supposed to diversify their portfolios to minimise the risk but were not to be involved in active monitoring of companies...The economists designing the whole scheme, however, intended IPFs to be future active owners of privatised companies who would play an important role in corporate governance. Thus a deep contradiction was built into the design of IPFs, a situation which lasted till 1996.[21]

The contradiction is simply another example of the confusion at the heart of the whole Klausian privatisation programme. The funds were the primary beneficiaries of the coupon process. They were to provide the link between the new private owners and the companies. And yet, legislation was in place which worked against the ability of the funds to exercise strong corporate governance. Much of the private property emanating from the coupon programme was thus effectively decoupled from the management which alone could restructure the companies.

(It is worth pointing out here that from an economist's point of view the asset stripping of companies and the corporate governance problems described above may not necessarily harm the economy in the long term. If a company manager transfers the plant's best machinery to a firm owned by his brother, for example, as long as it is used with a view to profit maximisation the economy as a whole will still benefit. A more pressing, though connected, problem which impacted on the economy and acted as a break to its development was the inability of all but a handful of very big, flagship enterprises to get hold of financing for investment. After an early period in which banks seemed to hand out loans to anyone who asked for them, it was not possible, or at least far too expensive, for small and medium sized business to finance expansion. The alternative, the capital markets, were simply dysfunctional for the reasons mentioned above.)

As if this were not enough, the relationship between the banks, the investment funds and the state provided a perverse logic to systemic failure which was as simple as it was devastating.

The investment funds to which the Czech people entrusted their shares stood under the control of the investment companies which established them. The investment companies, in their turn, were mainly established by the major financial institutions, especially the banks. The major shareholder in the major banks, however, was the state. In other words the dominance of the investment funds in the privatisation process immediately re-established substantial back door state control over the private property which the mass coupon method was supposed to have created.

It is unknown how often and to what extent this ownership line was actually used to influence the management of Czech companies. A more easily observable problem was the 'moral hazard' it inevitably created.

This moral hazard was the inner circle of a series of deficiencies within the conceptual apparatus of those who designed the Czech privatisation programme.

The weakness of the capital markets at the outer level drove companies to the banks for financing. When the companies, as some inevitably did, had problems repaying their debts, the banks found themselves faced with a conundrum. If they forced companies into liquidation they hurt their own investment companies, which controlled the funds which owned the companies.[22]

Neither could it have escaped the attention of banks with heavy state ownership that closing down large numbers of companies would, in the form of unemployment, entail a high price for the politicians who were their effective bosses. The result was bound to be inertia. The lingering threat of bankruptcy – the weapon of last resort hanging over all companies within a functioning market economy – was blunted into obscurity. The problem was compounded by Klaus's refusal to establish a functioning and effective bankruptcy law, hampering still further the already depleted arsenal of his country's new market economy. The figures in Table 5.2 do not tell the whole story. There is no indication of how big the companies were or of their importance in the respective economies. Many of the Czech companies may have been paper houses which had already been tunnelled. The broad trend however is clear.

A wholly contradictory alignment of forces was thus driving the Czech economy into the worst of all possible worlds: the banks had an institutionalised interest in taking the sting out of company restructuring and in so doing they were storing up a mountain of bad debt that would one day have to be paid.[23] As one economist put it:

Most of the time, indebtedness and loss making was not a serious threat to the survival of large firms. These firms could permanently spend more than they earned from their sales because big state-dominated banks provided credit freely...the banks tended to increase credit to avoid the large firms going bankrupt. This 'perverse' behavior produced many bad loans that could lead to a financial crisis in the future if corporate restructuring does not come soon.[24]

Table 5.2 Number of completed bankruptcies

	Czech Republic	Hungary	Poland
1990	–	–	29
1991	–	–	305
1992	5	1,302	910
1993	61	1,650	1,048
1994	290	1,241	1,030
1995	482	2,276	1,030
1996	725	3,007	984
Total	1,563	9,476	5,336

Source: EBRD transition report 1997, quoting Balczerowic et al. 1997.

Neo-liberal or politician without adjectives?

Before we come to the crisis which hit the Czech economy later in the decade, let us pause to consider the way in which we should interpret the philosophical underpinnings of the reform process as a whole. We have seen that Klaus and his team were bad neo-liberals. Would it not be more accurate to say that they were not, rhetoric aside, neo-liberals at all?[25]

It is difficult to pinpoint the exact time in the 1990s when journalists and foreign investors began to question the authenticity of Václav Klaus's commitment to New Right thinking. Klaus had, after all, the endorsement of Margaret Thatcher, he was an active member of the Mont Pélerin Society, he certainly talked as if he were in the New Right[26] camp and his opponents attacked him in terms of the rightist slant of his policies.

Klaus's reputation as a New Right politician was earned in the radical debate he had been instrumental in encouraging in the 'grey zone' Forecasting Institute in the 1980s. It was revealed to the world outside in the acrimonious disagreements he had had as Finance Minister in early 1990 with First Deputy Prime Minister For the Economy Valtr Komárek, the head of the same Forecasting Institute of which Klaus had been a senior member. Although Komárek accepted the need for a market economy he argued for a gradualist approach which he said was the only way to avoid economic chaos and massive industrial decline. He called for protectionist measures for major industries and opposed excessive fiscal austerity and monetary tightening. Against this, Klaus consistently argued that delay would be fatal.

There is absolutely no way to avoid the transformation shakeout of non-viable economic activities based on subsidized prices, artificial demand and sheltered markets. No amount of skillful macro-economic management, either through fiscal or monetary fine tuning, can prevent a GDP decline, an increase in unemployment, a once-and-for-all price jump after price deregulation, or a drastic devaluation before liberalization of foreign trade. Rational macro-economic policy can, however, avoid permanent galloping inflation, repeated devaluations, state budget deficits and growing foreign indebtedness.[27]

This was Klaus's claim to Margaret Thatcher's mantle: there was no alternative to the market whose logic could not be bucked. Governmental hocus pocus could do nothing to avoid the early pain of readjustment and from then on in it was, in Thatcher's famous phrase, all a matter of 'good housekeeping'.

But Klaus, at this stage at least, was more than all talk. In what could be termed the stabilisation phase, he sharply devalued the currency in three stages in 1990. A plan was formulated which translated into the liberalisation of most retail and wholesale prices from state control on January 1, 1991. Although a social safety net was guaranteed, the government's anti-inflationary macro-economic policy was the stuff of neo-liberal textbooks, focusing on restrictive budget deficits and tight money. The short term results were far less catastrophic than Komárek's prophecies of doom had led many to expect.[28]

Inflation rose from 10.8 per cent in 1990 to 56.6 per cent in 1991. Gross domestic product growth slowed from −1.2 per cent to −11.5 per cent. Industrial output fell 3.5 per cent in 1990 and 22.3 per cent a year later. From a western perspective the figures looked bad enough, but other countries in the region fared far worse. In the same two years, Poland recorded GDP declines of 11.6 and 7.0 per cent respectively. Inflation skyrocketed. In 1990 it hit 585.8 per cent, slowing to 70.3 per cent a year later. Industrial production was down 24.2 per cent in 1990 and another 8.0 per cent in 1991. A similar though less extreme scenario applied to Hungary.

Czechoslovakia's shock therapy was therefore a good deal less electrifying than that undergone by other leading reform countries in the region.[29]

This first stage of the reform process was probably Klaus's finest moment and he deserves genuine credit for the courage he showed in the face of faint hearts who would have postponed much of what was needed to an undetermined later date.[30]

With these achievements in the bag it appeared as though rhetoric and reality had achieved a perfect union. Klaus was the neo-liberal reformer he appeared to be. While all this was going on, a restitution law, returning property confiscated by the communists, had been passed in October 1990, a Small-Scale Privatisation Law, dealing with shops, bars, hotels and so on, was passed in November of the same year, and the Large Scale Privatisation Law was passed in February 1991.

What observers often failed to appreciate at the time was that Klaus the reformer was rapidly learning to be Klaus the politician.

> The genius of Klaus was to build a party that would provide solid parliamentary backing for economic reform and to implement policies that would win him consistent support, both in opinion polls and in a series of elections between 1990 and 1996. While Hungary and Poland also established strong parliamentary governments in 1990, these governments carried out deeply unpopular policies and therefore became victims of electoral backlash and internal fragmentation.[31]

Klaus and his circle of radical reformers, respected though they were, did not have things all their own way inside Civic Forum in the months following the revolution. We have noted the ideological differences with Komárek, but the more general objection, which Klaus clung on to throughout the decade, was that economic reform could not be left in the hands of dithering intellectuals, forever talking of the importance of a moral basis for the new prosperity and never actually getting down to the business of making anything happen. He was also anxious to outmanoeuvre the many émigré economists and would-be politicians that stood in his way. Tapping into these concerns, Klaus won over enough grassroots support to (unexpectedly) take the leadership of Civic Forum in October 1990. From here it did not take him long to realise that he would never gain a free hand inside the Forum and he took the rather unorthodox step for a party leader of creating a breakaway faction in February 1991 which established the Civic Democratic Party (ODS). At the 1992 elections, ODS, under Klaus's helmsmanship, established itself as the undisputed major force in Czech politics.

Some analysts have suggested that Klausian concessions to adversaries inside the first government created a form of 'social liberal' compromise, which as suggested above, prevented the kind of backlash against market reforms that would have unseated him had his New Right visions been translated into practice from the start. It also made it possible for people to vote for ODS without fearing for their security. The compromise took the form of maintaining much employment protection, some limited, albeit reluctant, cooperation with the trade unions, acceptance of the welfare system and the maintenance of price controls in socially sensitive areas such as energy, rent and telecommunications.

Those wishing to stress that Klaus was in fact a genuine neo-liberal see these moves as the necessary concessions that he had to make in a coalition government – the implication of course being that had he had things all his own way he would have acted differently.

Although it is impossible here to rise completely above the level of speculation, there seems little evidence to support this. On the contrary, social policy in the Czech Republic was marked by strong lines of continuity throughout the 1990s. Klaus did make moves in a vaguely neo-liberal direction, such as the withdrawal of some family benefits and increasing the age of pensionable retirement. He also spoke out against industrial policy and adopted a sharply Euro-sceptic tone in criticising EU social policy. But against this he said he opposed excessive income inequality, refused stubbornly to push through a workable bankruptcy law and even at the height of his powers never launched anything which could be reasonably described as an assault on welfare. He appeared to do everything to prevent unemployment biting, even though it was so often suggested to him that the economy was paying for a low jobless rate in terms of poor efficiency and laggardly restructuring.

So was Klaus really a closet social democrat? Those who want to make this case have some strong arguments in their favour which go beyond his consistent concessions to welfarism.

We have discussed the privatisation process in terms of serious deviations from neo-liberal precepts. That is to say, our frame of reference accepted Klaus's adherence to the neo-liberal agenda at face value and then sought to show up a mismatch between rhetoric and reality.

But important aspects of this process could also be interpreted as a classically social democratic way of privatising an economy.

The commanding heights – the banks, and many of the major enterprises – were of course kept under state control. The voucher method can be justified as a way of keeping economic power out of the hands of the former elite but it could also been seen in terms of a left-populist policy of redistributing wealth. The absence of foreigner participation also jelled well with the economic nationalism which so often typifies social democratic politics. There are additional aspects of Klaus's attitude to private property which smack of west European social democratic priorities. The housing market provides one clear example. Restitution may have given landlords the title to their buildings but this did not mean they could do what they wanted with them. Tenants could not be moved out without the provision of alternative accommodation. In regaining their properties, owners in most cases are required to respect the terms of existing agreements, meaning that rents are fixed not only below market rates but often below the costs of maintenance. 'Hence the property owning class in essence subsidizes the social safety net for the government...At times, restitution extracts from the propertied class in order to help support the population at large.'[32]

There is no question that Klaus made social welfare a top priority in his various incarnations as leader of the Czech economic reform process. It is no use saying that these were mere concessions although they may have started off as concessions in the first place. What after all is the cash value of a wholehearted commitment to neo-liberalism if it entails an endless series of compromises with social democracy? At what stage does this cease to be compromise and start to become part of the accepted agenda? And yet Klaus's rhetoric and the early reality suggests that he meant it, was saturated with neo-liberal ideology. He clearly showed a determined enthusiasm to liberalise prices, slash budget deficits and generally not to waste time in deference to the adjustment difficulties of the working man. He said he rejected industrial policy 'a priori'.This all leaves something of an untidy mess. Was he a macro-economic neo-liberal and a social welfarist at the same time? One possible way out is to incorporate the time dynamic hinted at above. Looking separately at the Klaus of the Forecasting Institute, the Klaus as finance minister up to 1992, his prime ministership up to 1997, and his chairmanship of parliament supporting the Social Democratic government of Miloš Zeman, we can identify four stages of development.

In stage one Klaus is a true neo-liberal intellectual. He has read the texts of Friedman, Hayek and other leading New Right thinkers. He has made contacts with these people and their supporters abroad. He knows the vocabulary and has learned what the main issues for neo-liberals are.

In stage two, Klaus enters politics and begins to push some of his ideas as policy. He may have a primitive understanding of markets and private property but this will neither be particularly evident nor hinder him in the tasks he initially sets out to accomplish. He will push neo-liberalism as an item in the policy debates of the early months of reform. He will provide the ideological energy for neo-liberal victory in the debate with Komárek and satisfy the aspirations of foreign governments and agencies in a series of early deregulations and fiscal balancing measures clearly flowing out of the neo-liberal agenda.

In stage three, Klaus assumes the premiership and negotiates the break-up of Czechoslovakia, a move which allows his new party to take the lead in policy formulation within a new sphere of influence. His party is the natural repository of right wing support in the Czech Republic but Klaus the ideologue must constantly fight for ground with Klaus the politician. Electoral interest groups are becoming more easy to identify and their call must be answered.

In stage four, Klaus is losing the political battle but mounts an amazing comeback and strikes a deal with a Social Democratic Party which a matter of weeks before he had denounced as representing a possible return to communism. By this stage, Klaus's commitment to neo-liberalism is in terminal decline. If it exists at all it is so heavily outweighed by the desire to retain a foothold in power as to make it an utterly insignificant part of his agenda. From Klaus's point of view, there have been so many concessions to social democratic policies already that it hardly seems worth making an issue of. He would rather settle scores with centre-right parties closer to him than the social democrats, than make a big issue about something as quaint as New Right ideology.

This four way split sees Klaus moving through the following incarnations: Klaus the reformer, Klaus the reformist politician, Klaus the politician/reformer and Klaus the purely opportunistic politician – a politician without adjectives one might say.

For those with a penchant for 'isms' this is not an attractive solution. The problem is that Klaus appears to have become a de facto social democrat from de jure neo-liberal beginnings. This is not simply because the ideological blueprints of the textbook can never be replicated in practice. Klaus increasingly showed little more than ceremonial attachment to radical neo-liberal reform. His inspiration as an ideologue was gradually replaced by his growing energy as a man of party politics. Whether he is or is not, deep down, still a neo-liberal is now beside the point. The point for Klaus is power.

Klausian economics coming home to roost

This chapter opened by questioning Klaus's judgement in having declared the transformation largely over by 1994. We saw how some of the key economic indicators were moving against the Czech Republic and we have examined the main ideological precepts, and the inadequacies thereof, at the centre of a decade of economic change. Let us now turn to an examination of just how deep the Czech Republic's economic problems are. How much better off, if at all, are the Czech people a decade after the economic reforms began? Is the massive pessimism described in Table 5.1 at the beginning of the chapter justified by the reality, and what are the prospects?

The appearance that years of economic reform in eastern Europe's former communist countries have produced widespread hardship for the many while showering the lucky few in mountains of riches sits uncomfortably with anyone who welcomed the fall of the Berlin Wall and the tearing down of the Soviet Union's empire in 1989. This perception, though, is widely shared and not only by commentators who talk of shock therapy and better times ahead. It is also greeted warmly by some as a 'told you so' riposte to the post-communist triumphalism of free market economists and politicians. Table 5.3 actually suggests that the best part of the last decade has taken the Czech economy backwards and not forwards. Taken as read, the gross domestic product figures show a clear and sometimes massive decline in real GDP in almost all the former communist countries

Table 5.3   Czech GDP in an international comparison, year 1998 (US$ billions)

	GDP ExR	GDP PPP	GDP % EU-15 = 100	GDP 1989=100
Czech Republic	55	133	62	95.3
Hungary	48	108	51	95.1
Poland	158	328	40	116.9
Slovakia	20	54	48	99.7
Slovenia	20	30	72	104.1
CEE 5	301	654	46	106.1
Bulgaria	13	42	24	66.8
Romania	38	134	28	76.2
CEE 7	352	829	40	97.2
Croatia	20	32	34	78.1
Russia	277	993	32	55.9
Ukraine	42	168	16	46.2
CEE 10	691	2022	31	69.4
EU-15	8331	7841	100	119.3
Germany	2097	1845	107	124.7
Austria	212	190	112	123.8
Greece	120	149	68	115.9
Portugal	104	148	71	126.7
Spain	550	641	78	122.1
USA	8219	8219	143	124.1
Japan	3965	3162	120	115.2

Notes: ExR = current exchange rate; PPP = purchasing power parity.
Source: OECD, Vienna Institute for International Studies (WIIW).

of eastern Europe and the Baltics. The Czech Republic has outperformed many but the results are hardly spectacular.

And yet the visitor returning to Prague ten years after the revolution would see a picture that the statistics do not paint: more high quality cars on the roads; shops brimming with western goods; people dressed in western clothes; restaurants offering foreign food; a city centre as wealthy looking in parts as anywhere in Europe. Which one of these alternative visions reflects reality? Is the wealthy exterior just that? Have some people indeed made big profits while the many have struggled to keep their heads above water?

The table can be seriously misleading since it is almost impossible to know how much weighting should be added to the post-reform economies to take account of the improvements in quality. Since communist economies were output driven – they had to meet plan targets – the incentive for enterprise managers was to cut corners on quality. Especially in the presence of new competition from abroad, shoddy goods were less and less acceptable to consumers. If an economy is made up only of cars, for instance, how does one take into account the change in the quality of Škoda autos now compared with 1989?[33]

It is hardly surprising, therefore, that the figures purporting to show large declines in GDP go hand in hand with some awkward companions. In the Czech Republic, during a period in which GDP ended up below the 1989 level, ownership of cars, washing machines, video recorders, microwave ovens and the like skyrocketed. In other words, the tangible results of strong gains in the standard of living seemed to be there. Private consumption, with the exception of 1991, was rising sharply throughout most of the decade.[34]

There were also social indicators that the real standard of living was on the rise. Life expectancy at birth rose to 70 years in 1995 from 68.1 in 1989 for men and to 76.9 from 75.4 for women. Infant mortality had fallen in the same period from 10.0 per 1,000 live births to 7.7. Secondary school enrolment had risen from 79.6 per cent of the relevant age group in 1990 to 97.4 per cent in 1995.

The consensus among economists appears to be that some progress has indeed been achieved in the transformation process but that many significant and painful steps were simply postponed. However, even if some of the more bearish analyses of the Czech economy's performance are based on suspect methodology, it can hardly be denied that things were getting worse rather than better in the last two years of the decade.

The warning signals were evident long before the country actually plunged into a recession presaged by two desperate looking reform packages within the space of two months in Spring 1997. The current account deficit was the biggest early signal that something was going wrong. From a state of near balance in 1994, a $1.4 billion current account deficit in 1995 had hit $4.3 billion in 1996. Two factors were responsible. The first was the crown-fort policy of the Klaus government, which was celebrated both as an anchor of the country's economic stability and as a trophy proclaiming the soundness of the overall economic reform programme. This allowed for a gradual real appreciation of the currency, making exports more expensive while popular foreign consumer goods became cheaper. The second, and more telling factor, was the underlying failure of company restructuring to provide domestically the goods consumers were driven to suck in from abroad.[35]

Not that this caused the government to change course or particularly worried many of its foreign admirers at the time. Since it could not be admitted that economic restructuring was not proceeding apace, the huge influx of imports was actually heralded in some quarters as evidence of economic success. The bulk of the imports were in fact investment goods which would be used to modernise industry. Latterly, the Klaus government was moved to impose an import surcharge. But this had more impact on Klaus's reputation as a neo-liberal reformer and came too late in the game to affect the trade deficit.

In the end, judgement was passed by the markets themselves. At the end of May 1997, massive pressure from international foreign exchange institutions forced the central bank to float the crown. It could be argued that the sharp fall of the currency was a completely unnecessary event as evidenced by its return to pre-devaluation levels several months later. A floating regime introduced two years earlier could have averted the crisis which was eventually brought on as Klaus's stubborn refusal to budge was held up to the markets as a red rag to a herd of bulls.

This was Klaus's most painful and public brush with reality to date. But it was not the first. The stock market had been in decline for years. By October 8, 1998, its lifetime low at the time of writing, the PX50 index had fallen 68 per cent to a mere 316 points from 1,002.4 on April 7, 1994. Although dealers and analysts were making loud noises about the weakness of the regulatory framework as early as 1995, few at that stage were ready to accept that the Czech Republic was anything other than a country which was bound to do well.[36] Another warning sign was the country's 'success' in keeping down the jobless figures. As late as 1995 unemployment was still under 3 per cent. In the same year, Poland's rate was 14.9 per cent, Hungary's 10.4 and Slovakia's 13.1. Was this, as the critics said, precisely because the Czech Republic had not undergone the restructuring that the others had? The answer was not quite that simple. For one thing, Czech unemployment probably looked better than it otherwise would have done because of the concentration of some of the Czechoslovak Federation's more appalling basket cases in Slovakia. Slovakia's contribution to the unemployment figures went with the dissolution of Czechoslovakia in 1993, providing the Czechs with something of a windfall gain in the jobless tables.[37] Nevertheless, the figures did look suspiciously low. They should at least have been taken as a signal. But once again, the warning was ignored. It was even argued that low unemployment was the fruit of the quick and efficient restructuring of industry that the Czech economic miracle had accomplished. Alternatively, the Czechs had cleverly managed to postpone unemployment until growth resumed later in the decade. As workers were eventually laid off, the dynamism of the economy would quickly put them back to work. Another chapter in the success story.

Why was everybody so optimistic about the Czech Republic and how were all of these signals ignored for so long?

Part of the answer is probably that the country simply looked destined for success. The capital exuded inherited architectural wealth. The setting was right. The Prime Minister spoke the New-Right language of the time with fluency and gusto. In Havel, the country appeared to have the embodiment of moral political authority itself. The two leaders may have disagreed on some issues but to most other reforming countries it still looked like an embarrassment of riches. Having bought into this image, the financial community took itself deeper with some extraordinarily naive methods of self-persuasion.

The most common refrain was that the Czech Republic was bound to do well because Czechoslovakia between the wars had been one of the richest countries in the world. Few articulated what this was quite supposed to mean in the 1990s, but the implication was that if they had done it once they could do it again. The communist years may have hit the work ethic but with a new economic environment the Czechs would soon get back to their wealth creating ways. All other things being equal, this would not have been entirely unreasonable. It is not easy to explain how economic knowledge and good practices are passed from one generation to another, but it is plausible at least to suggest that a country with a successful economic past to look back to for inspiration may have an advantage over a country which does not.

The problem, however, was that all other things were far from equal. The comparison was invalid because the make-up of a modern economy has changed fundamentally since the 1930s. Then the power house of a successful economy was its industrial base. A well educated, technically skilled workforce would just about suffice to yield success. Now, modern economies are dominated by the service sector. Strong technical skills and a good education are necessary but no longer sufficient. A successful service sector would depend on modern marketing methods, a devotion to the needs of the customer and an ability to sell in a highly competitive world.

How good were the Czechs going to be at these tasks? What happened before the Second World War was not going to provide the answers. But the questions were not generally being asked anyway.

Klaus's record of economic transformation is poor to say the least, and the disappointment has landed more egg on the faces of more economists, analysts, brokers, bankers and journalists than for any other transition country in the region.

When Klaus fell from power in 1997, appropriately enough on the back of a corruption scandal within his own party, some important steps in the right direction could finally be taken.

In April of 1998 a Securities Commission was established. Although it was not fully independent it could hardly have failed to be an improvement on what had gone before. Banking law amendments strengthened the central bank's powers to prise apart the relationships between banks, investment funds and enterprises. The bankruptcy law was amended in January 1998 to speed up liquidation procedures.

This was all to the good, but neither the interim government headed by central bank governor Tošovský nor the Social Democrat government of Miloš Zeman which succeeded it could do much to stop the roof falling in in 1998, the *annus horribilis* of the Czech transformation process. The year ended with economic growth falling by an astounding 3.9 per cent in the fourth quarter and 2.3 per cent for the year as a whole. The debt crisis had forced major banks into huge losses – Komerční Banka's 1998 loss was 9.55 billion crowns after a 455 million crown profit just 12 months earlier. Česká Spořitelna had not fared much better. Unemployment was also surging towards double figures.

Klaus himself has, unconvincingly, blamed the economic crisis which swept across the country at the end of the decade on central bank monetary policy and international financial institutions. A report by Standard and Poor's in November, which accompanied a credit rating cut, provided a more likely explanation and reads like an obituary for Klaus's credibility as an economic manager. The downgrade reflected the following concerns:

- Insufficient progress on restructuring the banking sector and many enterprises has left a relatively feeble base for economic growth, making the country susceptible to weakened exports, in the context of the slowing global economy;
- Piecemeal progress in creating an efficient legal framework to govern capital market and business conduct makes the Czech Republic more vulnerable than some other investment-grade transition economies to low foreign investor confidence; and
- The growing recession, following a few years of robust performance and increased private sector leverage, has raised public sector contingent liabilities at a time when the government's underlying budgetary position has deteriorated. With domestic credit at more than 70 per cent of gross domestic product (GDP) – the highest among transition economies – the country's financial sector is beset by a high and rising share of nonperforming loans.[38]

The country's initial advantages as a relatively low debt, high standard of living, industrialised economy have not been entirely squandered, and the S&P report praised the country's prudent macro-economic management, but ground has indeed been conceded to other leading reformist countries.

As Table 5.3 indicates, the Czech Republic still has a solid lead in the wealth stakes over Poland, Hungary and Slovakia. But the difference is not insurmountable. This realisation should serve as a warning to policy makers. The fact that the Czech Republic had already made so many mistakes on the way to the new millennium gives cause for a kind of muted optimism that certain lessons simply must have been learned. If not the consequences are clear enough.

How ironic it would be if at the end of the next decade of economic reform, the Czech Republic, once the first among equals in the post-communist world, found itself struggling to keep up with its neighbours rather than joining the rich man's club in the West whose membership it once thought it could claim by right.

6
Civilising Society

Establishing the form of a liberal-democratic capitalist system engaged Czech society at the level of the intellect. In the broad sense there was no great mystery over what needed to be done, although, as we have seen, this task has proved difficult enough. A more challenging problem came with infusing the new society with the cultural substance appropriate to a changed environment. The patterns of behaviour suited to a socialist, totalitarian system would need to be remoulded to fit new political, economic and national conditions.

This chapter looks at the transition from communism as it engaged the passions of the Czech people. Would the politicians and with them the citizens develop the prerequisite dispositions for a democratic environment? How would legitimacy be secured and more firmly rooted? In terms of the political parties and the success of regular elections this has already been touched upon. Here we are concerned more with a brief discussion of the way in which civil society has developed, how Czechs have come to terms with the past and whether the crucial ethic of equal concern and respect has established itself as a guiding principle for interaction between the country's ethnic groups.

Running parallel to all of this was the need to develop a civil economy in which capitalist forms were filled with business practices appropriate to their proper functioning. The market economy needed an enterprise culture. The business world would also need to incorporate new definitions of probity. Legal safeguards alone, crucially important as they are, must be complemented by a general acceptance that certain standards of conduct be respected. In the absence of general consensus on right and wrong, business dealings, supported at every turn by the law enforcement agencies, would be painfully slow. Policing mechanisms in the economy must be there and they must be efficient, but they should be called upon in exceptional circumstances only.

Corruption and the transformation

Since privatisation was one of the key political issues of the decade, the development of healthy political and business cultures was to some extent intertwined. This was especially true of the manner in which privatisation was conducted. Corrupt practices in this area could simultaneously destroy public confidence in the capitalist economic order and encourage profound mistrust of the politicians presiding over the transformation project as a whole.

An opinion poll[1] in October 1998 suggested that three-quarters of Czech citizens believed there was a lot of corruption among public officials and one in every four believed that almost all politicians and clerks were corrupt. Another poll earlier in the year found that over 50 per cent of the population felt they had been deceived by coupon privatisation while more than two-thirds thought it had benefited only the government and the dishonest.[2]

In view of what has just been said, this is hardly a matter of incidental importance. Corruption is dangerous in all political systems, but in a transitional period when loyalty to a new order of doing things is immature, the risks are heightened. Authoritarian forces can present themselves as the true friend of the small man and woman, promising to root out the selfish and immoral through the firm hand of disciplinarian control. In the Czech and Slovak Republics, the general argument has particular resonance. Patronage was perhaps the only legitimising weapon that the post-1968 hardline communist party had at its disposal. The sight of obedient party insiders getting their hands on all the material wealth while mouthing socialist slogans about the virtues of the working class provoked widespread revulsion throughout communist society. If the new boys after 1989 appeared to be getting up to the same sort of tricks, the conclusion that 'nothing has really changed' was all too tempting to draw.

Since people cannot be expected to volunteer the fact of their own misbehaviour to opinion pollsters, let alone the police, it is difficult to judge how widespread corruption has been. It could be argued that the incidence of corruption is in any case less significant than the perception thereof. There are, however, dangers in taking this line of thinking too far. Klaus for instance has frequently said that talk of corruption has been overblown. The danger here clearly lies in the risk of promoting the idea that the high level of corruption which is perceived to be going on is actually acceptable to the elite.

In general terms we have seen in Chapter 5 how the speed of privatisation was given priority over the institutional framework into which state enterprises were thrust. It was also argued that the absence of a solid legal framework prevented the emergence of 'private property' in any meaningful sense, allowing for a kind of free for all in which the slickest, quickest and dirtiest players could simply steal assets from an unsuspecting public.

If the official bias towards speed and against regulation made widespread corruption more possible, the sheer scale of the venture – around 40 per cent of state sell offs were conducted through the voucher method – made the temptations all the greater.

The biggest official scandal of the entire privatisation process centred around the case of Jaroslav Lizner, the boss of the coupon privatisation scheme and the head of the Central Securities Register.[3] The affair was significant for who Lizner was, what he did, how he tried to excuse himself, and perhaps above all, because he was actually tried, sentenced and jailed.

He was caught outside a Chinese restaurant in October 1994 with a briefcase containing 8.3 million crowns ($300,000) in cash. Lizner took the money as a bribe for helping the TransWorld International company (TWI) buy a stake in the Klatovské Mlékárny dairy from CS Fund. The high profile nature of the case inevitably provoked speculation on how many other senior public officials were engaged in such practices. Was corruption institutionalised? One of the most comprehensive studies of corruption in the privatisation process suggests that the kind of defence mounted by Lizner at his trial may have been instructive.

When Jaroslav Lizner was prosecuted for accepting a CZK 8.0 million bribe . . . the keystone of his defence was that the payment was not a bribe but a commission (later a deposit) for facilitating, as a private person, contact between two business entities. If we accept that this is simply untrue (as the court did) then an interesting question is raised: to what extent does such a defence simply reflect a natural strategy of self defence, in which the actor may be fully conscious of the corruption, and on the other hand to what extent does it reflect a failure of the individual in question to recognise his behaviour as corrupt . . . the nature of Lizner's defence indicates that the second factor may be in play, which in turn suggests that corruption may be more or less normal.[4]

Lizner was sentenced to seven years (reduced to six on appeal) on charges of accepting a bribe and abusing his office.[5] In an interview with Mladá Fronta Dnes in 1996[6] he alleged that hundreds of such deals had been conducted during the course of privatisation, and though, he conceded, this may have been unethical he pointedly refused to accept that he had broken any laws. Speaking of his own case he told the newspaper: 'We are not talking about morality, but law. And from this point of view the verdict was unjust and nonsensical.'

Corruption among officials in the privatisation process may have contributed to a sense of deep unease at the way some people were getting wealthy. Since the coupon method of selling state property was deeply intertwined with Václav Klaus's political personality – it was his distinctive selling point in the post-communist era – it was likely to have some impact on perceptions of politicians as well.

A more direct relationship between politics and corruption was to be found in party financing itself. Again, this matter should be seen in the context of building loyalties in a transitional environment. What sort of standards would the new parties establish in a country with no recent traditions of pluralistic democracy and what sort of message would they thus send to the people?

The most worrying thing about dubious party funding in the Czech Republic is that it has been clearly associated with ODS, the leading party over the decade and the party of the country's most successful party politician to date.

The most infamous early episode of absolutely open corruption involving ODS came with a party fund raising dinner on November 28, 1994. 'Business' men and women were invited to purchase tickets for the dinner for the sum of 100,000 crowns each. Given that many of the companies invited were fully or partly state owned[7] this meant that public money was being siphoned into a private political party's purse.

Soon after, the law was tightened up to prevent such flagrant abuses of power. But the affair, and especially ODS's refusal to acknowledge any wrong doing, could only contribute to a sense that those in positions of authority were participants in a free for all.

Of equal significance was the way in which corruption and allegations thereof appeared to have had no noticeable effect on ODS voter loyalty or the position of the party leader. It will be recalled that the catalyst for Klaus's fall from power in 1997 was a party funding scandal. Press reports had charged ODS with having 170 million crowns of illegal money stashed in a Swiss bank account. It was subsequently alleged that the money had come in connection with the 1995 privatisation of a 27 per cent stake in SPT Telecom to a Swiss-Dutch consortium, Telsource. Neither charge was proved. But the connection between the corruption in privatisation and party politics was thrust into the public consciousness once more. Senior ODS figures including Jan Ruml and Finance Minister Ivan Pilip used the occasion to try and unseat Klaus as party leader, and on this occasion Klaus took the bold step of asking Deloitte & Touche to carry out an audit and even admitted past 'mistakes'.[8]

Polls indicated that in the early months of 1998, support for ODS had indeed been hit by the affair although Klaus's massive last minute comeback at the general elections in the middle of that year suggest that the faithful were not inclined to exact significant punishment. All this was in spite of the presence of the Freedom Union, another right wing party with a similar policy stance and which took a strong stand against corruption. Loyalty to Klaus's personality, which was pushed ever more to the forefront of ODS politics in 1998, outweighed his voters' concerns about corruption. He remained at the helm of his party.

It wasn't only ODS of course. ODA was also tarnished by obscure party financing sources and, uniquely, appeared to pay the price, all but disintegrating as a result.[9] KDU-ČSL has been under a cloud of suspicion over dubious tenders at the defence ministry involving information systems for the Czech army. ČSSD was hit by the so-called Bamberg Affair in March 1998, which alleged that a loan from a group of Czech-Swiss businessmen had been accepted in return for the promise of government posts. At the time of writing the affair has not been resolved, having been effectively shelved. Another scandal broke after April 1997, when the tax office froze party accounts because of tax arrears. Help came to the rescue in the form of an 8 million crown loan from an initially anonymous donor who later turned out to be the co-owner of the Chemapol Group chemical giant. It was also reported in the press that one of the party's biggest donors in 1998 had connections with a company involved in the notorious tunnelling of CS Fund.

All that we have discussed above poses considerable danger to a fragile post-communist polity. Style is largely a matter of imitation and the responsibility of those at the top is clearly to lead by example. If medium-ranking officials see corruption established as normal practice above them, it is likely that many will feel inclined to practice it themselves. It is only a short step from here before the entire bureaucracy is affected, leading, sooner or later, to the institutionalisation of corrupt practices. Once such a culture is established it is also hard to root out since it has translated into expectations in terms of standard of living. Officials will tend to feel demoralised by and not cooperate with campaigns against corruption. Ending malpractice is, in this sense, welcomed in the manner of a pay cut.

If the politicians and the officials in the post-communist era will not abide by the rules of the game instinctively, does a liberal democratic society have any way of changing the culture from below? Many observers of east European politics have suggested that it does, although the method by which this is supposed to happen is not easy to define in concrete terms.

Civil society

The term 'civil society' has an abstract ring to it, making it the perfect weapon of the rhetorician – try pinning someone down on what exactly they mean by the term – and its scholarly connotations can give the shallowest of texts the appearance of profundity.

But civil society, according to some, is the core issue of post-communist political transformation. What then does it mean and how does it acquire importance in a liberal democratic environment?

The standard explanation of civil society is to describe it in terms of the intermediary institutions which fall between the apparatus of state power and the individual. Intermediary institutions can be anything from chess clubs to private newspapers or election monitoring organisations. This is all right in so far as it goes, but it tells us little. Let us take the first example of a chess club. Since it has the least obvious connection with politics, success in illuminating what is at stake should make the case all the more strongly.

Suppose a group of enthusiasts in a small town decides establish a chess club. They get together one Thursday evening and decide what to do. It quickly becomes clear that they will have to establish a set of rules. But if they are going to have rules they need a way of establishing them and people to enforce them. That is to say they need a constitution and an executive body.

In the communist system, the club would have been required to join up with existing organisations fully integrated, at least at the senior level, within the nomenklatura system. In something as innocuous as a chess club the ideological aims of the party would not be obviously felt. Lower level organs would generally not even be subjected to the usual political vetting procedures. But the possibility of participating in something genuinely independent and non-governmental would be compromised at an essential level by the knowledge of a connecting line running from the communist dominated higher echelons of the organisation down to the lowliest and smallest village club.

The sense of acting on one's own and joining up with others only on the basis of consent was conspicuously absent. Even as a member of a chess club the citizen was reminded of his place in a hierarchy of value and political power dominated by the communist party.[10]

Under a liberal democratic setup matters are, or should be, qualitatively different. There is no longer a procedural link between those who run the state and those who run the club. The metaphysical chain is cut. People are thrown back on their own resources. Constitutional and executive functions take on an exclusively voluntary character. Consensus unbounded and untouched by party ideology guides behaviour. It is precisely in this setting that individuals learn to abide by the rules of the game defining a liberal democratic political system. In electing officials they mimic behaviour in national ballots. But more than this, certain standards of behaviour become instinctively known to them as acceptable. Club members demand open and responsible dealings. They learn to expect probity in the running of an organisation with which they identify. People occupy positions of power because that is the will of the members and not because of favours granted or as a result of acquiescence in the aims of a dominant political ideology. Citizens learn and internalise modes of behaviour in the playground of non-governmental organisations which prepare them for graduation as citizens at the level of party politics in a pluralistic environment. It becomes natural to them that the standards of procedure they have internalised in the microcosmic environment of a voluntary association should be repeated on the grander scale of high politics.[11]

The problem for post-communist countries is how to bridge the gap and get to a situation in which this 'bringing up' process is sufficiently developed and widespread to affect the expectations of society as a whole and with it the political superstructure. Clearly there is a time and a generational element to be reckoned with. Those who have spent most of their lives in a totalitarian environment where such non-governmental organisations are absent cannot be expected to suddenly cast off habits learned over decades.

There is also the issue of changing the individuals who run the civic organisations. Members of associations whose higher bodies were once dominated by communists will not feel the changeover to a new regime at an instinctive level until those people have either been replaced or are accepted solely for their professional competence. Precisely because the communist system was so pervasive, involving itself in the running of the clubs and societies which are far more immediately understood at the popular level than parliamentary politics, individuals have a measuring stick as to how far society has changed even in the pursuit of their recreational activities. If members of the old elite are seen to have retained their little fiefdoms in the form of executive power over civic organisations, the perception of genuine change will be less than complete.

Understanding the importance of civil society at this level is easier than judging how much progress has been made. In the absence of any truly comprehensive research (at least to the knowledge of the author) evidence of change is almost of necessity anecdotal. Neither is it easy to see how government can do much to encourage civil society's development since its significance lies precisely in its voluntary character. As suggested in the introduction the purpose here is to raise the profile of civil society and make suggestions as to its significance rather than to draw conclusions.

What is a little easier to assess, and this may be indicative, is the degree to which society has come to a reckoning with members of the former elite. This is clearly important in so far as it sends out a message during the transition period (when day to day life for many people may not have changed significantly) that real change has occurred.

Recognition of the importance of drawing a line under the past came in July 1993 with the passing into law of a bill stating categorically that the communist regime was responsible for the repression of the past four decades and that it was both criminal in nature and illegitimate.[12] Even this largely symbolic statement was greeted with howls of protest by some, the Communist Party (KSČM) in particular. The lustration, or screening, law, was a matter of far deeper and wider concern.

Lustration, a word brought into the Czech language from the Latin *lustratio* (sacrificial purification), was designed to ensure that people who had had links with the state security police (StB) could not hold public office at least in the early years after the Velvet Revolution. Writing in 1992, Vojtěch Cepl, a Professor of Law at the Charles University in Prague, explained the motive for lustration thus.

> The emotional force behind the demand for widespread lustration results from anger about the invidious methods used by the former government to suppress all dissent, particularly after 1968...The secret police tried to infiltrate every school, office , and institution, particularly those employing intellectuals or involving any contact with foreigners. (Informal organisations such as reading or discussion groups were practically non-existent thanks to fear of police retaliation.) And since the borders were sealed, no one could escape the vigilant eyes and ears of the massive informer network.[13]

With such thoughts in mind, parliament passed a lustration bill with a five year time limit on October 4, 1991. Havel was uneasy about it but signed the law anyway. It was subsequently extended until the end of the year 2000, with a parliamentary vote overturning Havel's veto on October 18, 1995.

The significance of this to the civilisation of Czech society was obviously enormous. The StB files contained the names of around 140,000 people. But as one might guess by Havel's attitude, the matter was far from simple. The problem centred on the use of registers compiled by the communist secret police to 'convict' citizens of a democratic state. Since the essential charge against the last regime was its dishonesty, there was a clear irony in accepting documents compiled by its security police as authentic evidence of guilt. More specifically, collaborators were divided into three categories: A, B,and C. Categories A and B covered agents and conscious collaborators. But the third category, accounting for around half of the total, listed so-called 'candidates for collaboration'. The absurdity of using the latter category was highlighted even before the bill became law when Jan Kavan, a dissident and the current Czech Foreign Minister, was found to have appeared as a category C candidate for collaboration. His 'crime' was to have been in contact with an employee at the Czecho-slovak Embassy in London who was known to him as a member of the secret police. Kavan was eventually cleared (although it took until January 1996) but the damage had been done. Accusations against people such as Kavan not only undermined the credibility of the process, they were received with glee by the very people lustration was designed to expose. The 'look, they even say the dissidents were working for the StB' riposte fell to those who really were guilty like manna from heaven. The very nature of secret police activities was its secrecy. No one really knew who was involved. Catharsis was a neces-sary precondition for the readjustment of the national consciousness to the new society. The flavour of what came after communism would be determined by the willingness of the people as whole to come to terms with the past. The aim of constructing a new civil society would be hampered if the guilt of some and the bitterness of others could not find some point of release. And yet lustration showed just how difficult this precondition would be to meet. There was no easy answer. Jiřina Šiklová summed up the problem neatly at the end of a personal assessment of the lustration affair.

It is said that the old and blind Greek poet Homer died in a fit of anger when he could not solve a riddle posed to him by some children: 'What is it?' they asked, 'Those who we can catch, we kill and throw away. Those we cannot we carry with us.' The answer lice. In addition to being a children's riddle, the story has yet another dimension that could be applied to post-communist countries in their attempts to cope with the past. We can only cope with what we have grasped, identified and reflected upon. The unknown, that which we cannot grasp, we carry with us like lice and have no chance to shed until we grasp it. Too often, we simply forget that we carry this baggage, which nevertheless slows us down. We are like an obese person who forgets that he is loaded down with several dozen extra kilos and that these are an extra burden for his heart and ultimately shorten his life.[14]

Public opinion by the end of the decade was sharply divided. A poll[15] in February 1999 suggested that around 30 per cent of respondents wanted a new and better screening law and 14 per cent wanted the validity of the current law to be extended yet again. Conversely 20 per cent did not want an extension and 11 per cent wanted it abolished as soon as possible. A quarter had no opinion.

Outside the StB, very few officials who had committed violent crimes during the communist era were brought to trial. The most prominent, Miroslav Štěpán, the hardline Prague Communist Party leader, at the time of the Velvet Revolution was sentenced to two and a half years for ordering the police to attack demonstrators.[16] Alojz Lorenc, an StB general, escaped jail by claiming citizenship of Slovakia after the 1993 split. It was not until May 20, 1999 that two senior police officers were convicted in connection with the November 17 assault which sparked the downfall of the regime. But these were exceptions indeed. The vast majority of those who informed upon, imprisoned, tortured or killed innocents under communism walk among their fellow citizens unpunished and largely unknown. Almost none among the most senior members of the various communist governments have been convicted and jailed.

The media

A more readily accessible understanding of civil society comes at the level at which it makes itself articulate in the media.

Under communism, we should bear in mind, the media were not merely subject to the kind of censorship typical of dictatorial regimes throughout the twentieth century. It was not so much that there were limits on what could be said. The media were rather the propaganda arm of the communist party and charged with presenting events in the world in accordance with official ideology. They were in effect a branch of governmental authority. That is to say they were not really media at all. They could not act as a filter through which information moved up and down between the citizen and the state, altering and moulding the perceptions, ideas and, ultimately, actions of each.[17]

Czech leaders, Klaus included, have been generally aware of a need to respect the needs of the press in terms of holding regular news conferences at both party and governmental level. In contrast with neighbouring Slovakia, there has been no serious attempt by governments throughout the decade to pack the state media with loyal supporters or to harass critical journalists. There was some controversy over the removal of Pavel Šafr as editor of the *Telegraf* newspaper in 1994. *Telegraf* was owned by IPB Bank in which the state had a substantial stake.[18] It did not appear to be coincidental then that Šafr's ejection came in the wake of Telgraf's printing of several articles casting ODS in a bad light. But this does seem to have been very much the exception rather than the rule.

A more taxing problem was how journalists and editors would use the press themselves.

On the positive side, there has been a clear trend towards a greater willingness to criticise people in positions of power. The early days of consensus, when rightist politicians and businessmen could do no wrong, are long gone. Journalism is more aggressive, investigative and even better written. The most important points in straightforward reports are now more often than not found at the top rather than in the middle of a story. That is to say the idea of a developing news sense was becoming more clearly reflected in stronger, more focused leads. The media clearly regard issues of financial and political corruption as highly important and devote many inches of front page copy space to its reporting.

The practice of basing large numbers of stories on unnamed sources persists but is far less widespread than in the early years after communism. The happiest feature of the Czech media scene is that in contrast with many western countries, what is arguably the best general newspaper, *Mladá Fronta Dnes*, is also the most widely sold.

Television is a slightly different matter in that the headline-grabbing private Television Nova's main evening news has a strong lead in terms of viewership over state television, which adopts a more serious tone and tends to give less play to gory pictures of road accidents and more to foreign news items.

In general though, the news on the newsmakers is good. There is a strong commitment to serious news and the Czech people are fed with a daily diet of critical journalism on issues which matter in a liberal democratic state.

The most worrying side of the Czech media is that there is considerable evidence of corruption within its own ranks. This particularly involves the practice of companies 'buying' journalists to write complimentary articles about them. Proof is necessarily hard to come by since neither party to such deals is likely to shout about it, but everyone within the Czech media is aware that it goes on.

The impetus for such practices appears to come from both sides. *The Prague Business Journal* contacted several local journalists and public relations agencies in order to get a sense of just how widespread it is.

> the source of the problem comes from all sides...Journalists and cash-strapped publications, often looking for extra sources of revenue, often initiate the practice of article buying by soliciting money from companies they're writing articles about. And PR companies play along by acting as brokers for such 'paid articles...the director of the Association of Advertising Agencies said reporters often receive a press release, then call the company issuing the press release to say, "Okay, I read your press release. It looks good, but you should pay"'.[19]

This is not simply a disgrace to Czech journalism; it also threatens the reputation of one of the key institutions in a democratic environment. Respect for the press is not immediately forthcoming in a society where journalists under communism were nothing other than spokesmen for a discredited regime. Until the practice is stopped the integrity of the Czech media cannot be put beyond reproach.

Uncivil society

The issues highlighted above concern standards of behaviour within the Czech mainstream and attempts to swell the mainstream by understanding and breaking with the practices of the past. It isn't all bad news. Some of the problems of corruption at least are now being addressed. In April 1999 the government approved a draft amendment to the law on political parties, modifying the conditions under which gifts or sponsorship can be made. In the same month, the government moved to increase punishments for those offering and receiving bribes. Corruption generally is now a major political issue. This in itself suggests progress from the very early years after the collapse of communism. There can no longer be a sense in Czech society that abuse of public office or attempts by others to encourage this sort of activity is acceptable. Having gone through a kind of baptism of fire, the country may at last be moving forward. Civil society in terms of the voluntary organisations and the standards to be learned through them simply needs time to develop. This is the most frustrating aspect of the transition. Generations immersed in old habits will simply need to die out and be replaced by others whose frame of reference has been determined under different conditions.

There are, however, two very serious issues of concern to both the political and the economic sides to the transition, regarding standards of behaviour, which desperately need to be addressed if the Czech Republic is to develop into a successful liberal democratic, capitalist state.

Outside society

On November 8, 1997 a Sudanese student named Hassan Elamin Abdelradi was stabbed to death on his way from a nightclub by an 18–year-old skinhead who had chased him and a Sudanese colleague into a dormitory at the Economics University.

The attack was by no means the first racially motivated murder in the Czech Republic, but it provoked widespread anger throughout society and prompted public demonstrations around the country and calls in the lower house of parliament for a formal ban on the skinhead movement.[20] The killer was sentenced to 13 and a half years.

The murder of Hassan Elamin Abdelradi was only the most shocking and heavily publicised of dozens of racially motivated killings since the Velvet Revolution. Skinheads are a problem in other parts of Europe and their violent, aggressive ways are representative of only a tiny minority of the overall population.

But the tragedy of racially motivated violence should be set against the broader picture of deeply held and widespread hostility in Czech society against non-whites in general and the country's several-hundred-thousand strong Roma minority in particular.

An opinion poll at the end of 1996 found that just under 70 per cent of Czechs described their relationship with Roma as bad, 20 per cent were neutral and a mere 5 per cent said they were good.[21] Other polls showed more than a third favoured concentrating or isolating the Roma and nearly half favouring their expulsion from the country altogether.

Roma views on their own integration in Czech society were most graphically demonstrated after a single documentary on Nova Television in August 1997 showing Roma families living the good life in Canada prompted more than a thousand of them to pick up their bags and fly to Toronto asking for asylum.

Once the Canadian authorities moved to stem the flow, British air and sea ports became the next destination for the refugees. The exodus was evidence of a deep underlying problem.

In so far as we are concerned with the development of the Czech Republic as a country emerging from decades of communist rule, there is a clearly transitional factor at play here. The United States and Britain, for example, still have major racial problems of their own to contend with. But the ability to organise freely and discuss openly in the post-war era did at least allow the issue of racism against non-white groups to be addressed in the public domain. Few white Americans or Britons will now openly admit feelings of hostility to Blacks and Asians. Some may still harbour underlying prejudices and racially motivated attacks have by no means disappeared. But mainstream acceptance of racist attitudes is largely a thing of the past.

Evidence that things are rather different in the Czech Republic is available to anyone who chooses to raise the subject of the Roma in any Czech bar across the country. Events in the northern industrial Czech town of Ústí nad Labem in mid-1998 suggest that popular sentiment against Roma had in some quarters reached grotesque proportions.

On June 2 of that year a group of Roma handed the town mayor a petition against council plans to erect 4-metre-high walls to seal off two dilapidated housing blocks inhabited by 39 Gypsy families. At one point the mayor said they were simply noise blocks, but the fact that they were to be policed round the clock suggested otherwise. His own comments were revealing. 'This wall is not meant to separate people. It is not racially motivated. We simply want to separate the decent people from those that are not,' the mayor told reporters.[22] His definition of decency would be interesting to probe. But such remarks are generally typical of mainstream white opinion that Roma are lazy, dirty, and criminally inclined.

The plan to put Roma in a ghetto may have been the worst example of officially inspired racism, but it was in some respects less worrying than a highly discriminatory law which clearly sought to use the break-up of the Czechoslovak Federation in 1993 to make it harder for the Roma to acquire Czech citizenship.[23]

The problem centred around the many resident Roma who had moved to the Czech lands from Slovakia in the wake of the Beneš expulsions of ethnic Germans after the Second World War and under communism. The main issue was a clause requiring documentary proof of at least two years permanent residence and five years without a criminal record. Since the very exclusion of Roma from mainstream society had resulted in higher than average illiteracy rates, Roma were placed at an immediate disadvantage in dealing with officialdom. For similar reasons of exclusion, many Roma had lived in the Czech lands for years but had never registered as permanent residents. The criminal record requirement was in many ways the most spiteful. It corresponded to popular prejudice that the Roma were a lawless people who deserved what they got. Although studies suggest[24] that the crime rate among Roma is high, the type and pattern of crimes committed does not match that of the white majority. The disproportionate incidence of crimes such as pickpocketing and petty theft generally reflects the kind of criminality sociologists would expect from a group subjected to extremely high levels of unemployment and poor social circumstances generally.

This aspect of the law was waived in April 1996 but there was evidence that many lower level officials were not aware of the change. In any case the underlying requirement for Roma to prove their citizenship remained.

A letter to then Prime Minster Václav Klaus from the Commission for Security and Cooperation in Europe in April 1997 summarised the situation.

In recent weeks, news reports about your country have depicted thousands of anxious Roma families considering leaving the Czech Republic for refuge in Toronto, following the broadcast of a television program portraying Canada as a promised land for Roma. These people are clearly motivated by the prejudice, discrimination, and racist violence they face in the Czech Republic. We believe that the current Czech citizenship law significantly contributes to a climate of intolerance directed against this minority. We urge you to repeal the exclusionary elements of the Czech citizenship law as a concrete manifestation of your government's stated desire to integrate Roma more fully into Czech society.[25]

Racist behaviour is not only harmful to those who suffer it, although it must be borne in mind who the primary victims are. When it reaches the levels prevalent in the Czech Republic it also violates the liberal democratic commitment to equality before the law and the equal concern and respect for all citizens which officialdom must show if the whole of society is to feel fully rooted in its home environment. The failure of the far-right Republicans at the 1998 elections was a rare bright moment for the forces of anti-racism in the Czech Republic. But there is little evidence that this has been mirrored by radical changes in public opinion. It should also be noted that the racial problems we have seen in the 1990s have not come on the back of large scale unemployment or widespread social problems. As unemployment rises at the turn of the century this must put us on our guard against a resurgence of organised racist activity. In so far as this chapter has sought to identify the development of dispositions appropriate to a democratic system, racism is almost certainly the most worrying problem facing the Czech Republic. It brings dishonour on the country abroad, lowers the tone of political debate at the popular level and excludes an important section of the population from mainstream society.

Can't sell won't sell

Roughly speaking, the material ambitions of post-1989 Czechoslovakia were easy to identify. Politicians set their sights on bringing the country's living standards up to the European Union average. Many talked about a Czech path to the West. But in general there was no doubt what the end target was.

Ordinary people wanted to dress like westerners, live like westerners and be taken as westerners by the outside world. In order to achieve this, of course, they had to act like westerners. Above all they had to act like westerners in adopting the working practices that would give them the financial wherewithal to accomplish the task. The experiences of western expatriates, as businessmen and consumers, who arrived during the early 1990s and have remained, can therefore be used as uniquely well placed source material for observing the difference between what the Czechs aspired to be and what they actually were or became. They can serve as a yardstick. Not because that yardstick represents something which Czechs *should* be measured by but because they and their leaders, openly and tacitly, set this as the yardstick which they *wanted* to be measured by. If Czechs were achieving success in this project the expatriates at the end of the 1990s should have been seeing some drastic changes. The gaping difference between the kind of society that produces a western lifestyle and the one that wanted its benefits should have been narrowing. There are of course many different cultures in the West which support similarly high material conditions. The point is rather that there were core practices and attitudes which the Czechs needed to abandon.

The most obvious change was in their ability to market and then sell the goods which were going to give them the riches they wanted. The development of an enterprise and service culture was, above all else, the key cultural change that needed to be made. The western expatriates, as consumers, businessmen or journalists, carried around with them expectations forged in the kind of country the Czech Republic wanted to become.

The question of whether the Czechs can or cannot sell is far from irrelevant to their prospects of joining the frontrunners in western Europe. Buying and selling is after all what a market is, and a successful market economy is supported from below by a multiplicity of individuals and firms doing precisely that. Anyone of course can buy. The consumer driven import boom of the mid-1990s is testament to the fact that the Czechs were good at that. But what of the micro-economy, the bars, the shops, the service sector in general which make up such an over-

whelmingly significant proportion of a modern economy? If
Czechs did not re-orient their priorities away from themselves as
producers towards customers as buyers, how successful were they
likely to be? The spell of the producer oriented economy had to
be broken. The sullen, rude, slow, unhelpful, uncaring attitudes of
the communist era salesman, shopkeeper or service provider had to
go. At the very least, Czechs had to start seeing a relationship
between a contented customer and a healthy bank balance.
People weren't necessarily required, for the purposes of getting
wealthy, to be nicer. They just had to recognise that in a market
economy it is in their own interests to give the customer what he
wants, when he wants it and in the manner that he believes he
should get it.

Let us begin with what many would accept as the country's best,
certainly best loved, product: beer. The beer bar or Hospoda, is not
only the locus of much Czech social life outside the home, it is also a
core component of the country's service culture. The Czechs are the
world's biggest consumers per capita of beer. Pubs provide employment for tens of thousands and relaxation and liquid and solid
refreshment for millions.

Steve Misencik, an American of Slovak ancestry from Cleveland,
Ohio, remembers clearly his early experiences of service culture in
the Czech Hospoda. He first went to Prague in June 1990. The
practices he mentions were not universal but they were typical:
'Walk into a bar. Not showing menus, no prices, then forever saying
the item wasn't on the menu. Calculate the bill. Show the number
and then tear it up before you can examine it.' As far as the personal
skills of the waiters were concerned, Misencik's recollections are
uncontroversial.

'In the best case it's a cold, poker face. There is no small talk which a
lot of people are used to, especially in the States. That hasn't changed.
In the good places its solid service, but no chit chat unless you know
them.'

His most vivid recollection concerns an episode in a bar in the
centre of Prague's Old Town.

'Their tactic is to place a chair in the front door way with a sign in
English and German saying the place is full. Of course you can see
there are several places to sit down. We go in there to sit down. Didn't
see any menus. This goes back to the etiquette. If there isn't a menu

you should ask for what you want and you should know what you want. Don't waste the guy's time by asking for something they may not have. That will just annoy him. So we figure we'd better start off with some beers. In wander some Germans, order beers and ask the waiter, in German, if there was food. Despite the fact that others around, in clear view, were eating from large bowls and plates, the waiter said there was nothing to eat but that they have good food across the street. Sensing they have no card to play at the moment the Germans ask for the bill and thank him for the information and his kindness. He tells them 80 crowns for two beers. At this point we start wondering what our fate will be. In order to gain his respect we decide to drink two more beers. It came to 20 crowns a piece. The guy next to us paid 18.90. The waiter added his own tip which we would have given him anyway. We never did pluck up the courage to ask for a bowl of soup. This happened in the summer of 1998.' The last part of the tale is the most worrying – 1998. It has all the elements. Most bars are not as comprehensively appalling as the one described, though it is fair to say that over time an encounter with at least one such problem in the average bar is the rule rather than the exception. Having negotiated the first obstacle, getting through the front door, the customer was greeted with nothing that would have made him feel welcome. No readily available list of prices or products which would secure his position. The atmosphere was one of indifference. The customer had to pluck up courage, but didn't, to order food. Two foreigners were told to go somewhere else, presumably because the waiter couldn't be bothered to serve them. They were then over-charged. There is more, much more of this sort of evidence available. Few expatriates have not seen at least some of this behaviour and those that haven't were probably blinkered by linguistic inadequacy or simply tended to frequent higher class or foreign owned establishments where it was less likely to happen. Czechs themselves are generally only angered by such stories in so far as the impression is given that it is only foreigners who are thus treated.

The surly Czech waiter is the stuff of decades of communist and pre-communist society. But this is no reason to dismiss it as a cutesy aspect of the national character which locals and foreigners should learn to accept. It is imperative to realise that beyond the macro-economic models, the statistics, and the blackboards full of econo-metric equations, this is what an economy actually is. Owners, managers, and front line providers interacting with customers from home and abroad are the actors who give rise to the big economic picture most of us see in the economic reviews or the financial journals. They play a role from a script which is written deep into the national psyche and defines acceptable modes of behaviour. They at once imitate and then redefine the limits of what can and should be expected. Such forms of economic behaviour are not so much taught as imbued.

Evidence that the mentality described above does indeed spread far and wide into the economy comes from those expatriates who were working in the financial and trade sectors.

John Vax is a well known expatriate American investment banker who has observed business practices first hand throughout the decade. Vax starts at the most day to day level of corporate culture: 'How many times have you walked in to a Czech manager's office and found a copy of the *Harvard Business Review*? Never. If you go into a German, British or French manager's office you'll find it. They are up to date in the latest business thinking and the Czechs aren't.'

At the concrete level, Vax sees the same problem surfacing again and again: a failure to connect often well made products with the needs of customers. He quotes examples of two companies, whose names have been withheld, to illustrate the problem.

The first company makes machinery. Their competitors are German and Japanese companies and perhaps a Chinese company. The German and Japanese companies offer process control equipment, computerisation and a number of add ons such as manuals, support and warranties that provide customers comfort when they buy the product. The Czech and the Chinese just sell you the product, which is fine. But if they're only competing on price the customer here is going to choose the cheaper of the two, and that is the Chinese.

There is another company. They make thermal underwear and mountain climbing clothing. They have a patented process of making it that is superior, as far as I can see, to any western company. This is a start up company since 1989. They are smart. They are not competing on price. They say this is a better product so you pay for it. Everything is going fine there. Problem is that they don't offer the right colour, the right sizes, you'll have the bottoms but not the tops. They can't keep their clients supplied. So you can imagine that exporting is problem. They haven't a clue what they are up against. It only stems from the management. The owner has made some decent money but seems to be content and has not tried to allow the company to grow through establishing proper channels of marketing and getting the production process in harmony with the marketing. They'll run production of tops for three weeks and then switch to bottoms. They have good designers, good thinkers, they came up with a good product but now they can't bring it to the market.

Another expatriate businessman who did not want to be named for fear of offending his friends puts the point succinctly if cruelly. 'The Czechs can't sell. They can't even sell beer to you, their best product, without being sullen and often rude.'

If the Czechs are saddled with a mentality poorly fitted to the selling element of the buying and selling equation, which equals market economics in practice, the country has far deeper problems than many who sang the country's praises in the early 1990s had imagined. To say that the Russians are worse or the Hungarians are just as bad is beside the point. The Czechs after 1989 were no longer using the eastern countries as their measuring stick. They were looking to the West.

Many expats have been drawn to the conclusion that most of the improvements that have been made were concentrated in a layer of service culture industries and business which grew up on top of rather than from out of what went on before. While there are new restaurants, bars, and small businesses as customer oriented as their counterparts in the West, they are a distinct minority. The vast bulk of small and indeed big businesses hum along to much the same tune as they did at the beginning of the decade. For the numbers of enterprises which have made the jump to form the majority and then the norm, the old ones and the people who work in them will simply have to die off. It could take generations.

There is much in this section which is negative. But this is how the Czech economic and social changes looked from below to a resident foreign audience which was generally forgiving but nevertheless saturated with the kind of expectations which seemed natural to people brought up under the everyday economic arrangements prevailing in their countries. And it was the relative economic prosperity of those countries that the Czechs wanted to emulate. The methodology is of course anecdotal and descriptive rather than scientific. Economists often shy away from important questions about underlying attitudes and practices precisely because they are so difficult to measure. Market research based sample polls are probably close to useless since they cannot bridge the gulf between the way people say they should behave and the way they actually do behave. It is also difficult to suggest how the problems identified can be addressed. Perhaps real privatisation, real competition, the genuine threat of bankruptcy, market discipline, would help push things in the right direction.

To conclude on a dissonant if slightly tongue-in-cheek note, one further thought is worth considering. Do Czechs really want to make the necessary adjustments? Why should they? Nothing said above has been designed for didactic purposes. We have simply seen the juxtaposition of the Czech people's stated economic ambition and some observations on the ability of a sample of the real existing microeconomy to achieve it. The life of the average Czech citizen is far better than the dismal science of comparative macroeconomic statistics suggests. Most Praguers have country cottages or the equivalent which they visit most weekends through the summer. Only Londoners or New Yorkers of the highest income brackets have the same. Public transport is cheap, efficient, clean and safe. Violent crime is low. Cultural standards, despite the best efforts of Nova television, are extremely high. Many Czechs sneer at the tinseltown movie productions of the United States. (Though Hollywood blockbusters are always the biggest attractions at the cinemas.) The intellectual is respected and intellectual life is strong. Entertainment outside the home is generally cheap and readily available. Society is not ridden with the class based snobbery of some west European countries such as Britain. Community life appears to be solid, backed up by firm local identity, folklore and tradition. The country's central European location makes day trips to Germany, Austria, Poland and Slovakia an easy option. The Czech countryside itself affords financial and geographical accessibility to mountain skiing in the winter and excellent hiking in the summer. Czechs are respected in the world. They are neither poor nor rich but have most of the material possessions that all but the biggest victims of the western consumer society would want. There are of course many underlying social problems, just as there are in the West. The Czech Republic is no utopia, but life is good. If waiters are surly and business practices are archaic this is a small price to pay when set against all the advantages which exist alongside. In any case, hardly anyone can remember an enterprise culture that was better. Why should Czechs step down from the top of their own ladder to stand at the foot of someone else's? Could it be that the overall quality of life for the average Czech is such that there simply is not the incentive to drastically change lifestyles in the way that is necessary to produce western style prosperity?

7
The Velvet Divorce

Slovakia achieved its independence on January 1, 1993 as Czechoslovakia split apart in a 'velvet divorce' brokered by the prime ministers of the two republics, Václav Klaus and Vladimír Mečiar, the summer before. In the years that followed, the Czech Republic affirmed its reputation as one of the most stable and democratic states in the former Soviet bloc. Slovakia achieved the opposite. As Mečiar led his people into a twilight zone between democracy and dictatorship the country was increasingly shunned as a pariah. What went wrong?

The first obstacle Slovakia confronted was a clear tendency among foreign journalists and politicians to see the entire move to independence as the work of troublemakers. US Vice-President Dan Quayle visited eastern Slovakia in June 1991 bringing an uncompromising message that it was in the interests of both Slovakia and the region that the federation remain intact. Some of the more influential journalists were far less diplomatic: 'The break-up of the Czechoslovak federation is a sad unnecessary event that in the long run may benefit some sectional interests of the Czech economy and the *irrational fantasies* of some Slovak nationalists but is of little value to Central Europe' (my italics).[1]

Aspects of this initial opposition to the very idea of an independent Slovak state have endured in one form or another in both journalistic and academic writing ever since. The reasons are complex but take us to the heart of the main problems Slovakia has faced since 1993.

In this chapter we look at the key issues surrounding the break-up of the federation, starting with a crucial background discussion of the wartime fascist state of Father Jozef Tiso, whose tempting but ultimately false association with the cause of Slovak independence in the 1990s has proved a complicating factor in explaining the very good reasons for the break-up of Czechoslovakia in 1993. In the next chapter we move to the present with an examination of developments in Slovakia since independence that centres on the 1998 elections.

Disengaging from the fascist past

When nationalists in Slovakia looked for evidence that they were capable of managing their own affairs without guidance from the Czechs they found it in one worryingly obvious place.

On March 14, 1939, fresh from a meeting with Hitler in Berlin, Father Jozef Tiso addressed the Slovak Provincial Assembly. A motion was proposed and accepted. Slovakia became an 'independent' state for the first time and Czechoslovakia disappeared from the map of Europe. The importance of the wartime Slovak state to nationalists in the 1990s seemed obvious enough.

The regime may not have been palatable to everybody, but it protected the country and provided the conditions for the expression of national aspirations. Moreover, under most difficult circumstances, the Slovaks proved that they were able to govern themselves.[2]

A governmental administration had been established, the economy had developed, universities had expanded, cultural life had benefited generally.[3] There was of course the troublesome matter of Slovakia's role as a vassal state of the Third Reich, systematic anti-Semitism and participation in the German war effort. The history books would have to be re-written.

The most appalling example of precisely that came to prominence in 1996 with a book by Milan S. Ďurica, a professor of central and eastern European history at the University of Padua in Italy. The book was circulated to all Slovak primary schools on instruction from the ministry of education at the end of 1996 after 80,000 Ecu of PHARE programme money had been unwittingly granted to help finance the project.

The most comprehensive statement of what was wrong with Ďurica's book came in a letter to the Education Minister from the history section of the Slovak Academy of Sciences.[4] After pointing out that more than a quarter of Ďurica's work had been devoted to the six years of Tiso rule, almost as much as to developments in the nearly nineteen centuries up to 1848, the authors catalogue a sample of his errors and misinterpretations. The most salient points refer to Ďurica's treatment of the ruthless persecution of Slovak Jews under the Tiso regime and the deportation of tens of thousands of them to Nazi death camps. The text is worth quoting at length.

'He (Ďurica) describes the Slovak leadership, and especially President Tiso, as the saviours of the Jewish population, when in fact they were directly responsible for the deportations. In his efforts to exonerate him, Ďurica stresses that President Tiso did not sign the so-called Jewish Codex, even though it is well-known (although perhaps not to Ďurica?) that President Tiso never signed any governmental decrees. He does refer to some of the reports of the Sicherheitsdienst, in which the Germans indicated their dissatisfaction with the Slovak government's handling of the Jewish question. Yet, he completely omits any mention of the German regime's positive assessment of the Slovak leadership's performance in this regard. Nor does he mention Tiso's anti-Jewish speeches ... Most repulsive of all is his portrayal of daily life in the Jewish work camps (p. 162) as one of gaiety and happiness. Here, he literally claims that Jewish doctors in the work camps were paid for their services in gold when it was not available to common Slovak citizens. In Ďurica's account, it appears that to have been a Jew in Slovakia was in fact a stroke of good fortune.'

The Ďurica book was designed as a supplement for teachers of 10-year-olds. Not that this initially concerned Mečiar's Movement for a Democratic Slovakia (HZDS).

The HZDS has been attentively following the campaign, initiated with hatred and arrogance, against the university Professor, Milan S. Ďurica ... This book should become an educational supplement for public school students. In no case can the historical truth about Slovakia and the Slovaks be undermined or concealed. The HZDS has the deepest respect for everything that professor Ďurica has done for Slovakia and its well being and for making Slovakia more visible abroad.[5]

The book was eventually withdrawn but initial support from the ruling party provides an alarmingly clear picture of the kind of company some nationalists in Slovakia were keeping. Drawing inspiration from a fascist dictatorship was bound to colour the expression of Slovak nationalism in a democratic environment. Right from the outset, the nationalist consciousness was tainted by what it obviously perceived as a need to distort and deceive. The point is difficult to prove but it seems plausible to suggest that some of the aberrations of the Mečiar years could be traced back to this initial flaw.

A far more sophisticated attempt to 'save' Tiso in the 1990s came in a book by another Slovak academic living abroad, Stanislav J. Kirschbaum. *A History of Slovakia, The Struggle for Survival*, does not deny the fact of what Kirschbaum openly refers to as the Jewish Tragedy.[6] Instead he attempts to draw a clear distinction between the ideologically committed fascist prime minister, Vojtech Tuka, and Tiso, who is portrayed making concessions, against his will, to demands from anti-Semites at home and in Germany. Tiso is cast as man in a difficult situation. He must pay due respect to the overwhelming power of the Third Reich but he never loses sight of his role as guardian of the Slovak national spirit. In keeping with the title of Kirschbaum's book, Tiso, a man in struggle, does all he can to ensure the survival of his state. The message is clear: Slovaks in the 1990s can still be proud of this admirable man's achievements. The link with the wartime regime is sustained.

It is necessary to expose this argument once and for all not only as a matter of integrity but, as suggested at the beginning of the chapter, because its persistence simply gets in the way of discussion of the core issues. Nationalists themselves must also be persuaded that the Tiso regime was in fact a betrayal of Slovak national interests and not an affirmation of them.

The Tiso/Tuka dichotomy is in one sense true but trivial, and in another simply misses the most important point.

The sense in which it is true is that Kirschbaum correctly identifies Tuka as a man more completely imbued with enthusiasm for fascist ideology than his president. But different levels of adherence to official ideology are apparent in all regimes, and this also applied in the higher echelons of the Third Reich. There were undoubtedly high ranking officials in the Nazi leadership more immediately concerned with territorial expansion than others directly involved in racial genocide. Just as, of course, some officials were charged with road building, others with arms production and so on. But the distinction is obviously trivial in so far as they were all part of a regime with a specific set of purposes. They all contributed to the organic whole which they had chosen to be part of. Tiso may not have been as vicious an anti-Semite (although he made several anti-Semitic speeches) as Tuka, or indeed Hitler, but it can hardly have escaped his attention that joining up with the Nazis entailed support for and implementation of Nazi ideology. Tiso faced the same three-fold set of choices that all Europeans in Nazi occupied Europe faced: active resistance, passivity (keeping one's head down and hoping no one would notice), or outright collaboration. Tiso chose the latter.

The other major problem with the Kirschbaum argument is that it misses the point and does so in two spectacular ways. In so far as Slovakia under Tiso was 'forced' to accept official anti-Semitism including deportations, it had completely undermined its alleged status as an autonomous Slovak state representing the embodiment of Slovak national consciousness. Understanding this point is crucial as it fatally damages the Kirschbaum argument that compromise was necessary as a matter of 'survival'. (The word survival is used to evoke pity. Who could blame anyone for simply wanting to survive?)

Kirschbaum describes Tiso in the following terms:

> Basing himself on Slovak thinkers of the preceding century, he saw the nation as having a purpose in God's scheme of things. He accepted a hierarchy of values in the world, beginning with the individual, the family, and finally the nation. He rejected the existence of a Czechoslovak nation because he did not accept the notion that Slovakia and the Slovak people were just a topographical unit, an economic or cultural entity. He was persuaded that the Slovaks would eventually achieve statehood.[7]

With this motivation in mind, Tiso's role as head of a Nazi supervised state is interpreted as a series of unwelcome compromises to preserve the development of Slovak national consciousness. It is only possible to understand why this argument is so deeply flawed by recognising that anti-Semitism was the core ideological tenet of Nazism. This is what Nazism was and without it there would have been nothing we could recognise as Nazism. To concede, if indeed this is the appropriate word, Hitler's demands on this issue was not merely to accept, unwillingly or not, a mere aspect of Nazism; it was to infuse the Slovak state with Nazism's essence. Tiso as 'Vodca' replaced 'God's scheme of things' in Slovakia with Hitler's. The hierarchy of values, which Kirschbaum says motivated Tiso's nationalism, starting with the individual, moving through families and blessed by God, was replaced as the life blood of the state by a hierarchy of values defined by Hitler. Completing a bad argument, Kirschbaum says of the collaboration charge at Tiso's post-war trial: 'The collaboration accusation was ideological. Anything connected with fascism, in whatever form and for whatever reason, had to be denounced. The past had to be forgotten and eradicated.'[8]

Tiso is close to being portrayed here as an innocent victim. Kirschbaum simply fails to see that although Tiso's primary crime was collaboration in the murder of Slovak Jews (he doesn't even see that), in presiding over a state imbued with the Nazi will he had also allowed the jackboot of the Third Reich to crush Slovak national consciousness as his state's spiritual essence. The argument also suffers from more than a hint of anachronism. What if the Nazi's had won the war? When Tiso allied himself with Hitler it was by no means clear that Germany would be defeated. The Third Reich was meant to last for a 1,000 years. Tiso's collaboration was a once and for all measure. What is presented as an achievement in shaking off domination by the Czechs – fellow Slavs speaking a similar language – was rather the subjugation of Slovaks to an empire which viewed them as an inferior people. It is therefore a travesty to present the Tiso state as a step up for the Slovak people. They had in fact gone backwards to a hitherto unknown level of racially inspired servility.

As long as Slovak nationalists in the 1990s clung on to the Tiso regime as a legitimising device for their demands, it was probably inevitable that they were going to discredit the whole argument that the move for independence represented the development of a rising national consciousness. As we will see in some of the structuralist arguments used to explain the break-up of Czechoslovakia, this may also have had the effect of building something of an anti-independence bias into some of the more sophisticated academic texts.

It is perfectly possible to read the 1993 break-up of Czechoslovakia as the concluding event in a growing sense of national identity in

Slovakia and a recognition of fatally divergent priorities in the Czech and Slovak Republics.

The preamble to the Slovak constitution, which to the consternation of national minorities begins 'We the Slovak People'[9] and not 'We the citizens of Slovakia' as originally planned, speaks of centuries of struggle for 'national existence, and statehood'. As we have seen, to some this is mere irrational fantasy. In Prague in 1993 this opinion was shared by many ordinary Czechs and there was understandable nervousness among Slovaks in the early days that they had made a potentially fatal leap in the dark.

To those who opposed the split, opinion poll evidence provided prima facie, although probably superficial and misleading, evidence that neither Czechs nor Slovaks supported separation. Had a referendum been called, the argument goes, the perfectly viable Czechoslovak federation would have continued with two closely related peoples offering up a velvet coated counter-argument to the bloodshed accompanying small minded nationalism in Yugoslavia.

Before coming to the central issues which explained the mechanics of the split – to some the mechanics explain the split – it is necessary to take a very brief look at some of the evidence of this rising Slovak national consciousness. Contrary to the impression given in some writing and most talk at the time,[10] calls for independence did not of course arise out of a vacuum.

When exactly a national consciousness acquires the critical mass required to produce statehood is difficult to assess. What is absolutely clear about the 1993 separation is that *Czechoslovak* national consciousness had been edging into a state of meltdown ever since the end of communism in 1989. The problems go much deeper.

The essential difference between the Czech and Slovak nations was that modernisation had taken place at different times and under different conditions.

The time of Czech modernisation (once, long ago during the Austro-Hungarian empire, pre-1918) and of Slovak modernisation is the objective difference. Slovak modernisation proceeded and culminated under the protection of a paternalistic state, in this sense a social, not socialist, state, so that it did not become accustomed to the sacrifices which regularly accompany every modernisation or, indeed, all progress. In the process of the transformation to a political democracy and market economy it understandably found itself without the necessary resistance and showed certain signs of a weakened immune system for facing the temptation of populism.[11]

Modernisation here refers to the processes of industrialisation, urbanisation, literacy, demography, political pluralism and secularisation. In this respect some have noted as an irony the fact that the two nations shared far more in common at the time they split apart than when they came together. This 'snap shot' approach of course ignores, or insuffuciently recognises, the fact that character is formed as part of a process. The experiences on the way to modernisation, what had been seen, felt, endured, were the issue.

We saw in Chapter 1 how Czechs matured under the relatively benign domination of Vienna which allowed for increased assimilation into the structures of state power while Slovaks wrestled with the intolerance and intransigence of Hungary.

For Slovak nationalists, the early to mid 1800s are crucial years. Ludovít Štúr and others made decisive contributions to upgrading Slovak from a peasant language, frequently described as a mere dialect of Czech, to the status of an independent, codified modern European language. The nineteenth century as a whole represented a period of Slovak national awakening when the people acquired a better defined sense of their own identity complete with a developing literature and a flowering of cultural organisations. Although unacceptable to their rulers in Budapest, Slovak national demands in the 1800s are indicative of how far Slovak national consciousness had come. In the Demands of the Slovak Nation in 1848 and later in the Memorandum of the Slovak Nation, Slovak cultural leaders pressed for semi-independent statehood within the framework of Hungarian rule. After the 1867 compromise between the empire's Austrian and Hungarian masters the possibility of movement of any significant kind was blocked. Nevertheless, the aspiration for some form of statehood had been aired.

With the formation of Czechoslovakia in 1918, the Slovaks were catapulted from one form of domination to another, although the change in form was of course all important.[12] Czechoslovakism, which meant acceptance of Czech priorities and values, was substituted for Magyarisation – the change was to a far more benign and comprehensible form of subjugation, but it was a form of subjugation none the less. The Slovak agenda played a distinctly secondary role to the Czech and Czechoslovak agenda set in Prague. Although there were sound practical reasons for the two nations to come together – the minority issue mentioned in Chapter 1 being among the most powerful – the sense in which each accepted the new arrangements was defined by national identities formed in different conditions.

The point was neatly summed up by Jan Rychlík:

> though both nations welcomed the new state, each had different conceptions of it. From the Czech point of view the new state was the climax of the Czech national liberation effort, and for many Czechoslovakia was the restored ancient Bohemian state, whose independence had been lost in 1620... From the Slovak point of view the situation was completely different. In the new state the old programme for autonomy which had not been realised within Hungary was now to be implemented. In other words Slovakia was to be loosely connected with the Czech lands... For the majority of Czechs 'Czech' and 'Czechoslovak' were more or less identical and this was not acceptable for the Slovaks.[13]

Although, as we have seen, the so-called independent Slovak state during the Second World War represented a betrayal rather than an affirmation of the Slovak nation, the years between 1918 and its establishment saw the gradual development of many trends suggesting the upward movement of national consciousness. The most obvious was the rise of political parties with nationalist ambitions, most notably Andrej Hlinka's Slovak People's Party. Also, educational opportunities under the new regime increased the use of Slovak as a literary language translating into the emergence of a larger middle class with ambitions in the state apparatus. The very existence of the Tiso state, and the conditions under which it arose, make it impossible to say where this was actually heading. At the very least, it is clear that Slovakia was developing many of the pre-requisites to statehood, although there is no way of telling whether this had in fact translated into significant support for statehood as such.

Let us put a little more flesh on the bones here with a brief look at two of the key moments since the war.

The May 1946 elections provide the clearest evidence in the early post-war era of widely differing priorities among the Czech and Slovak peoples. The most obvious point of departure came with the enormous difference in the scale of support for the Communist Party in the two countries partly reflecting differing levels of industrialisation. In the state as a whole the communists gained 38 per cent of the overall vote. In the Czech lands this figure rises to just over 40 per cent but drops to a little over 30 per cent in Slovakia. Even more tellingly the communists emerged as the largest single Czech party but were beaten into a poor second place by the Democratic party in Slovakia which took 62 per cent. The communists, and everyone else, could hardly fail to draw the obvious conclusions. While the Czechs had given communism a vote of moderately strong support, Slovakia had overwhelmingly rejected it.

After the coup the Slovak national question was of sufficient concern to the party leadership in Prague to encourage a campaign against senior Slovak communists. Accused of bourgeois nationalism, the most famous victim, Gustáv Husák, was sent to jail and did not reemerge until the early 1960s.

In the more relaxed years after 1960, pressure from within the Slovak Communist Party concluded in the federal constitution implemented in 1969, an arrangement which at least nominally survived the end of the Dubček reforms following the Warsaw Pact invasion.[14]

The national councils gradually withered on the vine after 1969. But even if it had remained fully operational, the federation would in any case have been rendered impotent by Leninist party principles. Under the organisational system known as democratic centralism all levels of communist authority were subjected to central party control. The party in Slovakia, whose members led the governmental organs of power, was required to follow a line decided by the party in Prague. The appearance of federation in Czechoslovakia, as in the Soviet Union itself, concealed a fully centralised reality. Nevertheless, Czechoslovakia would hardly have gone to the trouble of setting up even the appearance of a federal arrangement without compelling underlying reasons for doing so. As we will see, structuralist arguments which focus on the internal logic of the move to independence from the point of view of the changed significance of the federal arrangements after communism often fail to give sufficient weight to this point.

Slovak nationalism after 1989

Inauguration of some aspects of federation was as far as Slovak national consciousness could go in institutional terms under a communist system. However, the 1989 revolution left the federation intact while removing the obstacle to its proper functioning. Many unanswered questions would now have to be resolved. One of the first, which typically many non-Slovaks regarded as trivial, was the country's name. In what became known as the 'hyphen war', Slovak deputies, reinvigorating a debate begun in the First Republic, attempted to achieve nominal parity with the Czechs by inserting a hyphen so that the country could be called Czecho-Slovakia. Slovakia would thus cease to be an uncapitalised appendage in the federal state's name. The refusal of Czech deputies to accept this provoked demonstrations in Bratislava, accompanied by calls for Havel's resignation, and eventually resulted in a compromise whereby the country could in fact be called 'Czecho-Slovakia' in Slovakia but 'Czechoslovakia' in the Czech lands and abroad.

This reluctance even to allow Slovakia the status of a capital letter was bound to enflame sensitivities. To the foreign media and the Czech people the whole affair looked absurdly small minded – Slovak small mindedness for proposing it not Czech small mindedness for failing to immediately accept it. The crucial point is of course that no one in Slovakia would have made an issue out of something as apparently innocent as a punctuation mark had it not reflected something much more substantial. If Czechs and Slovaks were having problems agreeing on a name it should have been clear that there were going to be bigger difficulties with some of the more obviously controversial issues such as economic policy and the terms of the new constitution itself.

It was the combined effect of different approaches to both of these issues which eventually brought about the end of Czechoslovakia.

As a way into this event it is worth bringing forward an argument put by Abby Innes[15] interpreting the split in terms of the development of party politics in the post-communist era. The argument is illuminating both for its strengths and its weaknesses.

Innes acknowledges that there were multiple causes for the split but bases her opposition to the idea that Slovakia and the Czech Republic no longer shared a common political culture on what she calls

Table 7.1 Public opinion on the preferred form of Czech-slovak state
relationship (%)

	Unitary state	Federation	Confederation	Independence	Other/ don't know
June 1990					
Czech Republic	30	45	–	12	13
Slovakia	14	63	–	13	6
November 1991					
Czech Republic	39	30	4	5	22
Slovakia	20	26	27	14	13
March 1992					
Czech Republic	34	27	6	11	22
Slovakia	13	24	32	17	14

Source: Quoted in Carol Skalnik Leff's *The Czech and Slovak Republics: Nation vs State*.
Boulder: Westview Press, 1997.

'evidence that a significant majority of both populations supported some form of a common state' and also 'the limited success of separatist parties.'

For those Czechs and Slovaks who still feel a sense of betrayal at the 1993 separation this point is often the beginning and the end of the argument. In the absence of a referendum, public opinion in 1992 could be gauged only by the results of the June general election and opinion poll evidence. But only by a superficial reading of both of these can a convincing case against the separation be made.

Table 7.1 indicates that support for independence in Slovakia was thin. Neither is there a significant move in that direction over the two years covered by the polls.

Other trends, however, are rather clearer. Support for a unitary state, that is abolition of the federation (a move back to 1968?), had minority but solid support in the Czech Republic. On average, the three poll figures put this support at just over one-third of the Czech population. The other obvious finding is the truly dramatic decline in support for the federation. In Slovakia, this represented a drop from 63 per cent to under a quarter. Czech support had fallen from 45 to 27 per cent. The introduction of the idea of confederation, mooted most often by Mečiar's HZDS, was by March 1992 the least popular option in the Czech Republic and the most popular option in Slovakia.

In other words, if the polls were accurate, there was fundamental disagreement at the popular level on how to resolve the constitutional question. It was not the invention of mercenary politicians though they may well have exploited it. The Czechs tended more in the direction of centralism, while the Slovaks were going the other way. And yet support for outright independence remained stubbornly low. Part of the problem was the sheer range of alternatives being floated at the time. Polls are notoriously less reliable the more options respondents are asked to choose from. What for instance was the real difference between confederation and outright independence? Confederation was the most popular option in Slovakia by 1992 but hardly anyone could offer a clear idea of what it meant. Most politicians at the time referred to it as something much looser than federation, perhaps with both countries having separate presidents, representation abroad, economic policies and even, according to some, a separate central bank. This would have given Slovakia at least as much independence as most countries currently enjoy within the European Union. The questionable reliability of the polls precludes any serious playing of the numbers game here, and the adding up of the confederation figures and the independence figures (it is worth a thought though) to produce 49 per cent in favour of independence to all intents and purposes. Confederation on this reading could be interpreted as the independence minded truth that dare not speak its real name.[16]

Still another problem arises with the time dynamic contained within popular preferences. Christian Democrat (KDH) leader Jan Čarnogursky, who replaced Mečiar as premier from 1991 to the elections in 1992, told France's Libération newspaper in July 1991 that he wanted Slovakia to become an independent state within the European Union by the year 2000. How many of his supporters expressed a soft preference for federation or confederation now – perhaps worried less about independence per se than who would run an independent Slovakia – with a harder preference for independence later?

The possible combinations are endless. The safest way to read the polls is simply to stress that support for the federation was sharply on the wane in both republics and that the Czech and Slovak people had clearly divergent views on the direction in which the problem would be solved.

In such circumstances it is by no means clear that a referendum would have shown majority support for the continuance of the Czechoslovak state. Much would have depended on the range and the type of questions people were asked to choose from. What if they had been asked to choose between the status quo and the two single most popular options in both republics – confederation or a unitary state? Polls suggest this would have produced complete deadlock and nationalists would have been outraged that independence had not been an option. With that option included we return to the hopelessly vague list of four separate choices. The politicians would inevitably have been called upon to sort the mess out. Even before we look at the interests and opinions of the key parties involved it is clear that the only option which would have unambiguously ended the confusion was in fact independence. It was a no nonsense solution when everything else was blurred. Opinion polls carried out in later years confirm that once the deed had been accomplished there was no great outcry and the issue is, of course, completely off the agenda now.

A series of inconclusive meetings on constitutional issues in 1990 and 1991 had resulted in agreement that both republics would draw up constitutional proposals by the end of 1991. The Czechs missed the deadline completely and the Slovaks produced proposals which begged all the important questions. After some attempts to work out a deal in Milovy in February 1992 and a botched effort by Havel to try and override the whole affair by asking the Federal Assembly to grant him the right to call a referendum, attention turned to the elections in June.

Klaus's ODS and Mečiar's HZDS emerged as clear winners. Klaus, whose party took 30 per cent of the vote in the Czech parliament, fought the election on the platform he had built for himself since the revolution – radical free market economic reform. Mečiar, whose HZDS took 37 per cent in the Slovak parliament, had won by promising a softening of economic reform to his unemployment ridden country and a solution, though it was not clear which one, to the national question.

The election erected an insurmountable barrier against retaining a common state in so far as, given the fact of the other, both election victors possessed greater incentives to separate the state than to seek a state-maintaining compromise. In the Czech Republic, the status of the ODS had depended upon its reputation for competence and the continuing success of rapid economic reform – a reform threatened in several ways by the election of a decentralizing and gradualist government in Slovakia. In Slovakia, on the contrary, Mečiar's power base had grown from his appeal to those who were fearful of too harsh a transition. Given, in addition, the negative attitudes of ODS supporters toward the HZDS and *vice versa*, the damage to the popularity of either compromising party would have been considerable and possibly fatal to the party leaders.[17]

Splitting the state also made Klaus and Mečiar possible as leaders of states rather than important but secondary figures to the world renowned Czechoslovak president.[18]

The argument thus far is solid but it should not be held in isolation from the reasons why the constitutional problem arose in the first place, and why it would have presented insurmountable problems to whomever had been leading the talks whatever their ulterior motives or interests.

Innes argues that: 'The limited time available for the development of a competitive party system emerges at the root of many of the problems that contributed to the division of the state.'[19]

The sheer immaturity of party politics in the two to three years after the end of communism allowed politicians to exploit the constitutional issue for their own ends. But it could as easily be argued, and from within Innes' own scheme, that the limited time available for party development in Slovakia left the outright independence ticket in the hands of a nationalist movement run by what were widely perceived as extremists. It is eminently plausible to suggest that many more moderate voters tending towards independence may well have been put off from expressing this preference for fear of being tarred with the same brush. As we have seen, some nationalists drew inspiration from Tiso's dictatorship, others were fiercely hostile to the large Hungarian minority. This all stands quite apart from the fact that Slovaks, like any other nationality considering independence, would have been plagued by doubt and feelings of insecurity. Preferences against outright independence would inevitably be motivated by a fear of moving into uncharted waters. For some these fears would dissipate with time; for others they would go only once independence was accomplished.

The fact that independence served the interests of ODS and HZDS does not mean that it *only* or even *mainly* served their interests or that the independence ticket would not have been taken up by different parties in a more mature environment to solve the same underlying problem.

It is also important to note that the divergence in economic policy was not some issue which the nationalist cause simply latched onto. The economy is a crucial part of what the nation is. Radical economic reform was one appropriate policy platform for a country with the kind of economy the Czechs lived in. In Slovakia the contrary appeared to apply.[20] Economic policies were forged in response to different kinds of economies and therefore countries. Innes tentatively speculates that 'Had the programs for transition, particularly economic transition, been more flexible, greater accommodation of the national conflict would, perhaps, have been possible.'[21] The different national economic priorities in each cannot be disassociated so cleanly from other aspects of nationhood.

The argument that the break-up of Czechoslovakia can be partly attributed to the development of the party system must also be set within the broader constitutional situation which encouraged calls for greater autonomy and eventually independence.

There are three connected elements to this which focus on the structure of the federal parliament and its relationship to the national councils, the legal setting for constitutional discussions, and the powers of the presidency.

Czechoslovakia had a federal parliament and each republic had its own national assemblies or councils. Political parties fought for seats both in each republic and at the federal level. This in itself was bound to encourage the formation of distinctly Czech and distinctly Slovak parties which could contest elections at both levels. A Slovak party which fought the republican level elections would inevitably seek to address issues which were of specific interest to Slovak voters and the equal and opposite applied for a Czech party. The presence of national councils established the basis therefore for parties with priorities directed towards the respective national groups. But these same parties, whose characters had been formed at the national level, also ran for seats in the federal parliament. In so doing their sense of national distinctiveness was re-enforced still further by the competition they now faced from parties formed in a different republic. When HZDS, for example, fought for federal representation it not only had to contend with rivals from Slovakia claiming to be more faithful guardians of the Slovak national interest, it also confronted Czech parties with similar interests in their own republic. Federal politics would inevitably be played out with one eye firmly on constituencies at home. Also, a federal party would find itself outmanoeuvred in republican elections by parties which claimed to offer something specific to the republican constituency and party organisation would inevitably tend towards division. Separate campaigns would be needed and separate organisations in each republic would have to be established to conduct them. Apart from anything else this erected a financial barrier which in the cash strapped early years after communism was far from insignificant. Finally, voters in a young democracy unused to pluralistic politics could hardly be expected to form loyalties to parties contesting republican elections and then switch them to others at the federal level.

For federalists this was a bad start. Party politics was organised along firmly national lines. At the 1992 elections, only the communists succeeded at the federal level in both the Czech and Slovak Republics, and even they ran under separate banners.

Even in the absence of such troublesome beginnings the rules of the federal parliament made any agreement on constitutional change hugely problematic. The federal parliament was divided into a House of the People with 150 members and a House of the Nations with another 150. But the House of the Nations itself was divided into Czech and Slovak chambers with 75 deputies in each. In order to secure changes to the constitution a three-fifths majority was required in each of these three constituents of the federal assembly. That is to say just 31 deputies in the Slovak chamber of the House of the Nations could veto constitutional decisions agreed upon by the remaining 269 deputies. If every deputy in the parliament had come from the same village this would have made constitutional change difficult. In the presence of separate party organisations and a host of different views on how much autonomy Slovakia should have, deadlock was all but inevitable. Slovak deputies were presented with a situation in which they could press their demands to an extent which far outweighed the relative number of votes they could muster.

But difficulties with the communist constitutional legacy did not stop there. While recognising the problem of the three-fifths veto system, Allison K. Stanger[22] has argued that the principle of using the 1968 constitution, and communist legality generally, as the basis for transition to new constitutional arrangemnents posed wider difficulties. For one thing, Article One of the Law of the Federation, though plagued with linguistic ambiguity and contradiction, appeared to place sovereignty in the hands of the separate republics. This suggested that Slovakia and the Czech lands had a kind of ontological priority over the federation which stood above them. Constitutional negotiations eventually resulted in agreement that both national councils would ratify any federal constitution.

In retrospect, the endorsement of the idea that a strong federation is comprised of strong republics was a fateful step. Involving republican legislatures in the ratification of a federal constitution would likely be a recipe for disaster, even in an ethnically homogeneous state. No republican parliament is likely to authorize the curtailment of its own power, which by definition, any movement to a federal system entails.[23]

Allowing the republics to take such an important role in searching for agreement was also hindered by the fact that 'Czech and Slovak National Council representatives, unlike their federal counterparts, had no experience of working together.'[24]

With these instituonal barriers in mind, when it became clear that Czech deputies were not about to concede enough to find a compromise, was this not the moment for a strong head of state to step in?

To those who believe the break-up of the federation to have been a mistake, happening against the will of the people of each republic, the voting requirements of the old communist federation, designed for a period in which unquestioned communist party control would make deadlock an impossibility, represent a classic poisoned legacy for a country in transition.

Havel clearly thought so, and by the end of 1991 he had performed an about turn on his earlier suggestions that the formal powers of the president actually be limited still further than those he already possessed. The most important suggestion was his unsuccessful plea for the right to unilaterally call a referendum. Had he been able to appeal over the heads of the politicians, or so the argument goes, the president would have been able to hold the federation intact. This all depends greatly on how one reads popular sentiment at the time, which as we have seen was far from predictable.

In any case, overriding the parties carried with it serious dangers of its own. When Havel wanted to call a referendum in September 1991, senior figures in both the Slovak Christian Democrats and the Movement for a Democratic Slovakia were opposed to the idea. Had the Czechoslovak president been in a position to ignore these objections he would have been lunging headlong against the two most powerful forces in Slovak politics.

There were also other dangers. Some suggested that a referendum would have forced the mainstream Slovak parties to reach some sort of compromise and take a more moderate line in the negotiations. But by depriving them of the anti-federation ticket there was a serious risk of handing the initiative to more aggressively nationalistic groups such as the Slovak National Party.

On this reading it is possible, therefore, to see the presidency's limited powers as a blessing in disguise rather than evidence of institutional inadequacy. Intervention from on high by a more powerful Czechoslovak president risked leaving the biggest political parties in Czechoslovakia thoroughly dissatisfied and probably embittered without having solved the underlying problem.[25] In the end, the only thing Havel managed to secure from Slovak deputies was a refusal to support his own re-election as federal president. He could not even sustain himself in his own office never mind hold together the federation as a whole. This was the last straw.

By mid-1992, Slovakia and the Czech Republic had separate party systems, separate parliaments, different social problems, different economic policy priorities, no federal president, no big federal parties, contradictory popular opinions on constitutional arrangements, and two hard headed leaders with vested interests in a split. The centre could not and did not hold.

The argument that there were institutional and legal dynamics pulling Slovaks and Czechs apart is certainly helpful but it does not tell the whole story. Although this is much more obviously true in retrospect, the main driving force for a republic-specific party system was the presence of separate national councils existing underneath the federal structures, and conjoined with this a critical mass of cultural differences and policy divergences to feed these parties and make them viable. Should the national councils have been abolished in order to encourage Czechoslovak-wide parties which only contested federal elections? Unless one actually wanted to see molotov cocktails on the streets of Bratislava, the question once posed reveals itself as absurd. But it is useful at least in so far as it teaches us to avoid detaching the very existence of state institutions, and even their legal frameworks, from the consciousness which gives rise to them. The national councils existed, at least in the case of Slovakia, because of a republic-wide desire for national self-expression. These councils then gave rise to national rather than federal oriented parties. In conditions of free speech, it gradually became clearer that there was a lot to talk about which was different in each country. National consciousness at various times may have been inchoate, desultory, and fickle. But this is natural for any nation making tentative steps towards autonomy and subsequently independence.

Had the federal parliament been differently organised so that a three-fifths majority of the institution as a whole sufficed for constitutional agreement would matters have been different? Perhaps the least speculative way of answering this question is to say that matters would almost certainly not have been more peaceful, simple or clear cut. The lack of significant minorities from either country residing in the other *à la* Yugoslavia certainly contributed to the peaceful nature of the split. But an amicable separation was also made more certain because nationalist forces were not, in the main, radicalised by years of frustration and unsatisfactory compromise. There was no substantial welling up of national hatred because the institutions of the federal state acted as a safety valve, deflating the hard-line nationalist cause precisely by giving in to it.

After the end of communism, Czechoslovakia was clearly unviable as a joint state of Czechs and Slovaks in which both nationalities could live comfortably together. The federal parliament's constitutional voting requirements and the legal setting which surrounded it, designed under a communist system where there were no free votes or negotiations, probably brought matters to a head more quickly. But this was not the real issue. Differing economic and social problems lay on top of widely differing historical backgrounds. Czech and Slovak national councils, themselves the product of a recognition of distinctiveness, gave rise to parties which were encouraged to play on this. In the face of public confusion, redoubled by opposing ideas on a solution in either republic, the party politicians had to do something. It was in their own interests to split the state but their own interests should not be divorced from the separate nations which produced them.

When the federation disappeared it went with a whimper rather than a bang.[26] Some regretted the end of the common state but there were no mass demonstrations against its demise. Far greater passions would be aroused over the way Slovakia's leaders ran their newly independent state. It is to a discussion of this subject that we now turn.

8
Surviving Mečiar

For the first time in their history, Slovaks woke up on September 27, 1998 to find themselves living in an independent state with the real prospect of rule by a fully democratic government. The euphoria which greeted the crushing defeat of Vladimír Mečiar at the general elections at home was matched by praise from abroad. A massive 84 per cent of those eligible to vote had turned out at the polling stations, most with the aim of telling the world outside and the government at home that their aspirations were the same as their counterparts' in the rest of Europe. No band of crooks and cronies was going to rob the Slovak people of its birthright. The elections were hailed as an affirmation of the democratic spirit in the face of adversity, and so they were.

The primary beneficiaries of Mečiar's defeat were the Slovak people themselves but the significance of the elections to the rest of the region should not be underestimated. When US Secretary of State Madeleine Albright warned of Slovakia becoming a 'black hole' in the middle of Europe[1] she was not merely lamenting the country's democratic record. With a 500,000 strong population of ethnic Hungarian's and a location between Ukraine and Austria to the east and west and Poland, the Czech Republic and Hungary to the north and south, events in Slovakia had geopolitical significance as well.

With aberrations such as the politically motivated kidnapping of President Michal Kováč's son in mind, Slovakia had already been demoted from the first wave of European Union and NATO candidate countries, and official hostility to the Hungarians raised the prospect of ethnic disturbances at the very heart of Europe.

Western diplomats publicly observed protocol and maintained a steady if concerned silence. But privately many were clear about what was at stake. 'This is it,' said one in August. 'If Mečiar wins this one, they're finished. Forget about joining the EU and NATO and start thinking about Lukashenko's Belarus or even Milosevic.' It was a close run thing and those worst case theorists who talked of Mečiar rigging the elections had plenty of cause for concern as we will shortly see.

Explaining Slovak politics to a foreign audience is complicated because many of the main events of the past decade appear so bizarre that it is tempting to simply write the country off as one of those maverick places which will never be understood. Due deference is shown to that opinion in the restrained pace of the argument as it unfolds and the assumption that many readers will have relatively less knowledge of what has gone on here than in the Czech Republic. The aim is, however, to show through an understanding of the development of his political personality by repeated betrayal from erstwhile allies that there is a clear logic to be found in Mečiar's degradation of democracy in Slovakia. This chapter addresses that core issue in Slovak politics in the 1990s and focuses on the 1998 elections while moving back into the post-independence past and forward into a description of the new government, how we should understand it and what we should expect from it.

To anyone who remained to be convinced, the events surrounding the Mečiar government's illegal thwarting of a referendum on direct presidential elections in May 1997 showed just how far Slovakia had fallen since achieving independence four years earlier. But the fact of Mečiar's open violation of the constitution and disregard for the constitutional court was of more than didactic significance. This also marked the point at which the main forces impacting on Slovak society at home and abroad crystallised into distinct form. It was the last straw for the EU, which excluded Slovakia from early accession talks. The disparate domestic opposition began to join forces as never before. With Michal Kováč due to step down in March the following year and no prospect of a compromise candidate emerging from the country's bitterly divided parliament, Mečiar had ensured that he would add the powers of the presidency to those of the premiership in the run up to the elections. The stage was set.

The idea of a referendum on changing the method of selecting the president from a parliamentary ballot to a direct vote of the people flowed directly from a recognition of the implacably polarised nature of Slovak party politics as it had developed from the 1994 parliamentary elections.

At that time, Mečiar's victorious Movement for a Democratic Slovakia formed a coalition government with the far right Slovak National Party (SNS) and the far left Workers Party (ZRS). These parties represented but also helped perpetuate the extreme versions of Mečiar's own utterly populistic style of politics.

HZDS had been formed as a breakaway faction from Slovakia's answer to Civic Forum, the VPN (Public Against Violence) in 1991. In some respects there is a parallel here with Klaus's Civic Democratic Party in the Czech Republic. Both were formed out of a conglomerate party grouping a broad coalition of anti-communist forces and both were suspicious of the intellectuals which led them. In distinct contrast, however, HZDS had no special ideological programme to match Klaus's neo-liberalism. It was argued in Chapter 4 that the personality of the ODS leader was given special emphasis partly because of the difficulty of selling complicated political and economic ideas to a broad section of the public and partly because loyalty to individuals rather than parties was more readily comprehensible to voters in a post-communist environment. Nevertheless, the party itself had a clear set of principles with which its members could, if they chose, identify.

Lacking any substantial policy initiatives, HZDS at the party level focused heavily on the figure of Mečiar without the mediating influence of an ideology on whose wings other individuals could rise to prominence. This was reinforced by the formation of a voter base identifying itself strongly with the Mečiar leader figure and seemingly unwilling to part company with him however outrageous his behaviour became.

But Mečiar's influence did not stop there. It had equal significance in the evolution of the party political scene in Slovakia as a whole and, of course, on his country's style of government, with all that that entailed in the economic transformation, the development of democracy and foreign policy.

If Vladimír Mečiar's personality became the main issue in Slovak politics we need to pause for a clear discussion of how it arose in the form that it did.

Creating Vladimir Mečiar

Mečiar rose to prominence as Slovak prime minister in the republican government after elections in 1990. VPN was officially led from the outside by Fedor Gál and other intellectuals who had forged a broad ranging platform for democratic change. The top figures in the VPN considered their role as one of defining the character of post-communist politics rather than shaping it as politicians. This admirable if somewhat starry-eyed attitude – which may also have reflected a recognition of their own unpopularity – discouraged Gál and his coterie from running for office, clearing the way for the much more pragmatic and hard headed Mečiar while the party leadership focused its attention on federal politics in Prague.

Problems were not long in coming. The VPN party leadership became increasingly concerned at Mečiar's combative style, particularly his attitude to constitutional and economic change and rumours that he was using secret police files against opponents. After ignoring repeated warnings Mečiar was ousted as premier in April 1991 and driven into opposition.

If Mečiar was considered a divisive figure by some of his most prominent party colleagues, opinion polls putting his support at anything up to 80 per cent suggested that this view was not shared by the Slovak population as a whole.[2] VPN, a new political grouping with tenuous bonds of loyalty, had parted company with its most powerful political asset.

The consequences of dumping Mečiar were threefold. The Platform (later Movement) for a Democratic Slovakia, which had been formed as a faction within VPN became a separate party, HZDS, with immediate voter appeal centred around Mečiar himself and replacing VPN as an effective force in Slovak politics. In addition Mečiar had been taught to fear enemies from within his own ranks as much as from outside. He had good grounds for feeling aggrieved. It is genuinely abnormal for a premier to be ejected by his own people while at the height of his popularity. If approval ratings of the sort he was getting were not sufficient to keep him in power, democratic politics had been shown to be a much more complex game than it may have initially appeared. Mečiar had learned a hard lesson in power politics. Thirdly, having been pushed into opposition, Mečiar was encouraged to play the national card ever more strongly as a means to distinguish his party from the rest of the pack.

With this in mind and on a typically vague set of policy issues, promising to soften the impact of Prague inspired economic reforms and to defend Slovak interests in constitutional negotiations, HZDS walked away with the 1992 elections, taking 37 per cent of the vote in the Slovak National Council which after 1993 became the new country's sovereign parliament. The next biggest party, the Party of the Democratic Left, got just 15 per cent.

With a victory of this scale under his belt and formal support from the ultra right SNS, Mečiar took Slovakia to independence and simultaneously established himself as prime minister of an independent country.

Although he had been a latecomer to the idea of all out independence Mečiar could now present himself as the founder of the Slovak state. Moreover, the opposition's ambivalent stance on separation from the Czechs allowed him to frame the political debate in terms of support for or opposition to Slovakia itself. Standard party disagreements on policy were thus infused with a new urgency and the implicit and sometimes explicit accusation from Mečiar that his opponents were motivated by treasonous intent.

Events in 1994 sealed the development of a political personality that had already acquired some distinctly worrying traits.

Having once been ditched by erstwhile supporters he lost power a second time in March, to be replaced by a coalition headed by members of his own party.

His downfall was presaged by a series of defections from within HZDS which depleted his power base in parliament. He also lost the support of his coalition partner, the Slovak National Party (SNS), which split. The final blow came in the form of a blistering attack by the country's first president, Michal Kováč, who had himself been a high ranking member of HZDS.

Mečiar was back in power after elections later that year, but the lesson he had been given at his ousting in 1991 had been hammered home once again. He would never forget it.

Mečiar had shown that he had sufficient electoral support to make him the most powerful force in Slovak politics, but betrayal by erstwhile supporters, and in the form of Michal Kováč a president who owed his position to a parliament which Mečiar dominated, had prevented him from translating this into stable governmental power.[3] Having re-established himself after elections in late 1994, Mečiar would seek to enforce discipline within his own party, neutralise the defectors and discredit and deplete the powers of the presidency. The logic of all of this flowed directly from the experiences described above. Mečiar had been taught that anyone who was not explicitly with him was potentially against him. He would now seek to accomplish his aims regardless of the law, the constitution and the interests of Slovakia abroad.

After the 1994 elections, lengthy negotiations eventually brought the SNS back into Mečiar's fold. The SNS had been the only party to consistently advocate independence since the revolution, and having achieved that maximal aim in 1993, the party's nationalism turned away from the external enemy image of Czech domination and moved inwards towards the Gypsy and Hungarian minorities. Hostility towards both took on a distinctly racist character.[4] Since Mečiar depended on the SNS it was inevitable that he should make concessions to the constituency which supported it. His consciousness was thus tinged by a need to play to a gallery inhabited by people with a tenuous commitment to racial harmony in a country bordered by five others and which governed the interests of two large minority groups. As if that influence was not disturbing enough, the dubious HZDS–SNS duo was joined after the 1994 elections by the newly formed Workers Party (ZRS). It had broken away from the fully reformed communist Party of the Democratic Left (SDL)[5] and played hard on the most populist socialist instincts of a people brought up under communist rule. The forces of vulgar nationalism and uncompromisingly populist socialism thus buttressed Mečiar on either side.[6]

The first signs of what the country was in store for came with Mečiar's so-called 'night of the long knives' on November 3 and 4. Around 50 privatisation decisions were cancelled and the media and state security services were brought firmly under the control of Mečiar and his allies. The response to the elections themselves provided an even clearer signal. The Democratic Union,[7] which had been created by HZDS defectors and had helped unseat Mečiar in 1994, was accused of using forged petition lists of the 10,000 signatures needed to formally register as a party in the elections. This was a serious accusation. Mečiar was looking to overturn the mandate of a democratically elected opposition party, have its deputies ejected from parliament and recoup the bulk of their votes for himself. He failed to convince the constitutional court but the attempt was designed to intimidate and served as a warning that any future breakaway groupings from his party would be dealt with severely. The message was reinforced two years later when dissident HZDS deputy František Gaulieder was actually expelled from parliament.[8] Gaulieder had quit HZDS on 5 November 1996 citing anti-democratic tendencies within the party and shady privatisation deals. He also claimed that he was being followed by the security services (SIS).[9]

In keeping with developing HZDS practice of bringing forward only the flimsiest of pretexts for flagrant violations of the constitution, a forged document complete with a false signature was produced. The document purported to be a letter from Gaulieder himself asking to be relieved of his duties from parliament. He was. Mečiar had succeeded in expelling a democratically elected deputy from the Slovak parliament, and neither the constitutional court nor howls of protest from the domestic opposition, the European Union and the United States could stop him.

The Gaulieder affair had compromised the integrity of the country's parliamentary system and it brought Slovakia firmly into the twilight zone between democracy and dictatorship. The explosion of a bomb outside Gaulieder's home a month after leaving HZDS was a further warning that opponents of the government were in for a rough time.

To no one in Mečiar's Slovakia was this message brought home with greater clarity than the country's first president Michal Kováč, or more precisely his son of the same name, whose kidnapping in 1995 began what must rank as one of the most extraordinary series of events in post-communist eastern Europe.

The president had emerged after the 1994 elections as the only centre of constitutional authority challenging Mečiar's hegemony. Although his formal powers were limited, he served as a public base for the opposition and, of course, Mečiar's hostility was heightened by Kováč's past as a HZDS deputy.

In the wake of the 1994 elections Mečiar had moved quickly to slash Kováč's budget allocation and the SIS was set on his tail to uncover anything which could be used against him. The perfect weapon appeared to have been found with the revelation that the president's son had become embroiled in a corruption case involving a Slovak export import company, Technopol, and its dealings in Germany. The German authorities wanted to question Kováč junior as a witness in the case. The affair had the potential to cause acute embarrassment to the president.

On 31 August 1995, Kováč junior was kidnapped by masked men, forced to drink at least one, some reports said two, bottles of whisky, beaten up, given electric shocks, tied up, dumped in the back of a car, whisked across the border to neighbouring Austria and left outside a police station. The hope clearly was that the Austrian authorities would extradite him to Germany, where involvement in the ensuing Technopol fraud investigations raised the prospect of blackening the family name. The plan backfired badly. An Austrian court not only refused to have anything to do with the matter but openly suggested that the Slovak authorities had been behind the kidnapping.[10] The European Union and the United States both issued statements expressing deep concern.

The botched kidnapping showed up in bold relief how far Slovakia had moved away from standard democratic norms.[11] This was real criminality. The security services were clearly being used to serve the political interests of Mečiar's government and there appeared to be no lengths to which they would not go to achieve their aims. But there was worse to come. Robert Remiáš, a former police officer, had struck up a relationship with Oskar Fegyveres, a member of the SIS, who claimed to have been involved in the kidnapping. Remiáš was acting as a go-between for Fegyveres and the Slovak media. On April 30, 1996, his car exploded with him in it. No one knows who dit it.

The official version of both events was about as believable as the notion that Gaulieder had left parliament voluntarily. Kováč was said to have kidnapped himself – if indeed any kidnapping had happened at all – and Remiáš's car had exploded by itself as a result of faulty fuel injection equipment.

The question of whether Mečiar was directly involved in the kidnapping – it is quite possible that the SIS was acting on its own initiative – is of secondary importance. The political environment had been degraded so severely that organs of state power felt unconstrained by law or basic morality. Lawlessness was in danger of becoming routine.

The referendum

As long as Mečiar remained in power, there was no danger of serious investigation of any of the abuses of power cited above. But the 1998 parliamentary elections were looming. What would happen if Mečiar lost? Both he and especially his allies in the SIS could potentially find themselves in deep trouble. They needed an insurance policy. The powers of the president to grant amnesties, and so Mečiar thought, to prevent prosecutions in advance were a prize worth fighting for.

Since President Kováč was due to step down on 2 March, 1998, and parliament was too divided for any political grouping to muster the three-fifths majority required to produce a winning candidate, Mečiar knew that he could scoop up these presidential powers as they passed naturally to the prime minister.

Unfortunately for Mečiar the opposition knew this too. The Christian Democratic Movement (KDH)[12] had proposed a constitutional law on a directly elected presidency in December 1996. A change in the rules to allow for a direct vote of the people seemed a sensible way of avoiding the prospect of the country having no head of state at all, and HZDS on some occasions had been inclined to agree.

A petition was launched in January to raise the necessary 350,000 signatures to force a referendum on the issue. Matters were complicated by a proposal passed in parliament in February for another referendum containing three questions on NATO membership. But armed with the necessary signatures, President Kováč added the presidential question to the ballot papers so that all four issues could be settled at one go.

Mečiar had been outmanoeuvred, or so it seemed. A direct presidential election, which he knew he would lose, would deprive him of the possibility of granting amnesties to those involved in the activities over which he had presided since the 1994 elections. With the prospect of his losing the next parliamentary elections as well, investigations into past wrongdoings would inevitably be launched and who could say where they would lead?

Drastic measures were needed and they were taken. The Interior Ministry, ignoring the constitutional court – but by now who could have been surprised by that – solved the problem simply, unilaterally removing the question on the presidency from the referendum ballot papers.

Victory was again Mečiar's, but this time, and coming as merely the latest in a series of highly public scandals, the costs were high. The first major blow came with the European Union's decision to exclude Slovakia from the first wave of EU accession hopefuls. The importance of this in rallying support against Mečiar cannot be overstated. A newly independent, small European country which had emerged from decades of submersion in the Soviet bloc faced isolation once more. In view of the relative success of neighbouring Poland, Hungary and the Czech Republic, the EU decision stunned the population. This was a major public humiliation. The effect was felt with particular keenness among the young, more fully imbued with a sense that they had been born into a lucky generation and equally determined that the mistakes of the past would not be repeated.

The European Commission's report explaining Slovakia's demotion was a damning indictment of all that Mečiar had done to his country, the more so because of the diplomatic and understated style in which it was written.

The operation of Slovakia's institutions is characterised by the fact that the government does not sufficiently respect the powers devolved by the constitution to other bodies and that it too often disregards the rights of the opposition. The constant tension between the government and the President of the Republic is one example of this. Similarly, the way in which the government recently ignored the decisions of the Constitutional Court and the Central Referendum Commission on the occasion of the vote on 23/24 May 1997 directly threatened the stability of the institutions. The frequent refusal to involve the opposition in the operation of the institutions, particularly in respect of parliamentary control, reinforces this tendency. In this context, the use made by the government of the police and the secret services is worrying. Substantial efforts need to be made to ensure fuller independence of the judicial system, so that it can function in satisfactory conditions.[13]

The referendum campaign also served to inspire the opposition to put its differences aside and finally present a more united front against Mečiar's excesses. The most important initial step was the unification of the forces of the right which came with the creation of the Slovak Democratic Coalition (SDK).

The SDK, which gradually evolved out of the petition action for the referendum, grouped the two main Slovak rightist parties – the Democratic Union and KDH – along with three other much smaller groupings. Opinion polls put SDK support at anything up to 30 per cent. For the first time since the 1992 elections HZDS ceased to be the undisputed major force in Slovak politics.

At around the same time, HZDS itself was weakened by the departure of Mečiar's foreign minister Pavol Hamžík. Hamžík had never been a member of the party but he was yet another senior figure, who like every single one of Mečiar's foreign minister's before him, had found the prime minister's behaviour to be intolerable. Joining forces with the charismatic mayor of Košice and current president Rudolf Schuster, Hamzík also set about the formation of his own party, the Party of Civic Understanding (SOP), which served as a natural reservoir for disenchanted HZDS supporters.

The left came on board as well. The SDL, itself disgusted by Mečiar's behaviour, distanced itself ever further from its ambivalent 'third force' suggestions of the previous year.[14] Along with the ethnic Hungarian Coalition,[15] which could always be counted on to oppose Mečiar, the opposition was crystallising into a formidable bloc.

One other crucial element underpinning all of this was Mečiar's loss of the private media. From near total support at the time of his first ousting in 1991, the major newspapers, with the one exception of Slovenská Republika, had moved against him. State television and radio could be relied upon to be slavishly loyal but the gradual and initially careful granting of private licences to television and radio stations added a complicating factor to Mečiar's efforts at control.

The key event which turned the tables decisively against him came in March 1997. A meeting of striking actors and opposition deputies converged inside the Culture Ministry on SNP Square demanding to meet the minister. Instead of coming downstairs to address their grievances, the minister called the police to disperse them and left through a side entrance. Television Markíza, by far the most popular station in the country, carried pictures of parliamentarians and actors being jostled and manhandled outside a government ministry on its main television programmes. It was the beginning of a gradual move which brought the most powerful medium in the country firmly onto the side of the opposition. Mečiar's sense of anger at Markíza's turn against him was all the greater because the station had been put under the control of Pavol Rusko, a man who had started out as a faithful ally.

We have seen how important aspects of Mečiar's authoritarian style of government flowed directly out of a political personality formed through the treatment he had suffered at the hands of former colleagues.

The list of political abuses cited above is designed to be illustrative of a certain internal logic rather than a chaotic series of developments in a new post-communist state. This not only characterised Mečiar's government but shaped an opposition, including the private media, which could not consider working with him and united forcefully against him.

It should be recognised that the massive economic corruption which characterised his rule was also a product of this bipolar world of friends and enemies.

Coupon privatisation, which began under Klaus's auspices in Czechoslovakia, had been divided into two waves. The second, which went ahead after the 1993 split, was cancelled in Slovakia in favour of direct sales, often to enterprise management. The most notorious but highly indicative example of Mečiar's attitude to privatisation came in February and March 1994, around the time of a no confidence vote in parliament which brought his government down. On March 15, 27 direct sales were approved in one night. The most conspicuous beneficiary was Alexander Rezeš, who acquired a stake in the country's biggest company, VSZ a.s., of which he later became president. After Mečiar's resumption of power, later the same year, Rezeš, was appointed transport minister and in1998 he managed the HZDS election campaign.

The idea that enterprises could get into the hands of anti-Mečiar or, as bad, non-Mečiar, supporters, was anathema. This was not simply because of the potential for enhancing his party's financial situation. Corruption in Slovakia was not just institutionalised as in the Czech Republic; it was systemic.[16] And its systemic nature flowed directly from Mečiar's style of government and the world-view that inspired it. People who could not be relied upon in the business sphere would turn against him as had those who had betrayed him in politics.

The run-up to the 1998 elections

By the time President Michal Kováč stepped down in March 1998, Mečiar was in deep trouble. Support for his biggest coalition partner, the Workers Party, had collapsed. The bulk of the media was against him. Not surprisingly, he was shunned by the West. Opinion polls put support for the combined opposition at up to 65 per cent. If this translated into votes at the September general elections he would not only be defeated, the opposition would have a three-fifths majority, enough to change the constitution and reverse everything that Mečiar had done.

Predictably, and shamefully, Mečiar did use the presidential powers inherited from Kováč to grant amnesties to those involved in the Kováč kidnapping case and also to prevent prosecutions over the presidential referendum. The first was designed to protect his government's security forces, the second to protect his own interior ministry. Mečiar had moved to close the circle ahead of his impending demise.

But was this really enough? The financial, political and security police oligarchy he had built around him had a lot to lose and the opposition was not likely to be very forgiving in view of what it had been through in the previous four years.

It was not immediately obvious what the government could do. Scare tactics were one option. In early January, the government said it had received a telegram informing it of a probably foreign inspired plot to assassinate Mečiar by February 25. But stunts of this sort impressed no one. The state television media was another obvious candidate for abuse. But it was already seen as little more than a vehicle for HZDS propaganda and opposition supporters watched Markíza instead.

As the opinion polls refused to budge it quickly became clear that an assault on the election system itself was Mečiar's only hope.

In the midst of all that happened in post-independence Slovakia it is easy to lose sight of just how close the country came to a total breakdown of the democratic system in 1998.

On May 20, of that year, just five months before polling day, the Slovak parliament passed an election law changing the rules for coalition groupings by raising their threshold for parliamentary representation.[17] Since the only political subjects affected were the Slovak Democratic Coalition and the Hungarians, the law was a blatant attack on the opposition, and it occasioned yet more public criticism from the United States and the European Union. Mečiar had timed his attack perfectly. It was too late for the SDK and the Hungarians to seek redress in the constitutional court and soon enough before voting to threaten the opposition with disarray. The groupings under attack from the new law would have to change their statutes and join together as individual parties. They would thus be concentrating on internal matters in a crucial period before a general election.

Since the SDK was made up of five parties and the Hungarian Coalition of three, Mečiar may have been counting on the differences between each of the constituents to prevent the consolidation which alone could overcome the terms of the election law. But the strength of the opposition's hostility, taken up a notch or two by the law itself, was sufficient to encourage both affected groupings to pull together as individual parties.

Mečiar had one last try. He applied to the central electoral commission to have the SDK, his biggest and most dangerous rival, de-registered from the elections altogether. With many of his own supporters on the electoral commission there was a serious risk that he would win. In the event the forces of decency and democracy in Slovakia were just strong enough to stop him. But only just. In August, the electoral commission registered SDK by a vote of 18 to 17. The supreme court subsequently confirmed the decision. Had SDK been banned from the elections one can only speculate on what would have happened next. The opposition had threatened to bring its people out on the streets. And what then? Would Mečiar, as many in the opposition at the time suggested, have used civil unrest as a pretext for postponing the elections? One vote.

The election campaign itself was predictably based almost exclusively around Mečiar. Less predictable was HZDS's stunning success in garnering the services of a host of international megastars. Since the opposition was focusing heavily on accusations that Slovakia was isolated abroad, high profile appearances with Claudia Schiffer, Ornella Mutti, Claudia Cardinale, Gérard Depardieu, and when his father declined, Jean Paul Belmondo's son, constituted the perfect reply. The opposition was as stunned as everyone else.[18]

But the voters had made up their minds months ago and the opinion polls would not budge. The only potential problems the opposition faced were in mobilising their voters to actually go to the polling stations or the rigging of the elections. In answer to both problems, civil society provided its own solutions. One of the many ironies of the Mečiar years is that civil society under the very pressure that was exerted upon it may well have emerged stronger and healthier than in neighbouring countries with more normal democratic arrangements.

It was noted above that young people especially had been outraged by exclusion from the first group of EU candidate countries the year before. Many had already come to the conclusion that Mečiar's antidemocratic behaviour was taking their country down the wrong road. With demotion from the EU their ranks were swelled and feelings were radicalised further. It was from among such people especially that many of the civic organisations which played a role in the election campaign came.

Two of the most prominent in the latter stages were Občianske Oko (Civic Eye) – which was set up to monitor the election campaign and the vote count – and Rock the Vote, a collection of musicians which went around the country staging mass concerts to encourage people to go out and vote.

Andrej Salner, a 21-year-old Graduate of Brandeis University, was Občianske Oko's press spokesman. His background and his views are typical.

For many people it was make or break time. We felt a profound urge to make this country normal, liveable and democratic. Had the 1998 elections been compromised Slovakia would simply have ceased to be a genuine democracy. There was mistrust on both sides of the political spectrum. The opposition suspected the government of wanting to abuse the elections and the government even claimed the opposition wanted to do the same. There were foreign observers but they didn't know the situation so well and there weren't very many of them. Who was more competent to oversee the elections than the citizens themselves?[19]

As a parting intimidatory shot, there was a scare a few days before the vote when it was announced that the army would join the police outside the polling stations. But with the country, the media, and a host of voluntary organisations staking their all on a free vote, and

Table 8.1 Results of 1998 Slovak parliamentary elections

Parties	Per cent	Seats
Movement for a Democratic Slovakia (HZDS)	27.0	43
Slovak Democratic Coalition (SDK)	26.33	42
Party of the Democratic Left (SDL)	14.66	23
Hungarian Coalition Party (SMK)	9.12	15
Slovak National Party (SNS)	9.07	14
Party of Civic Understanding (SOP)	8.01	13

Source: Central Electoral Commission.

some, like Občianske Oko, deploying hundreds of monitors to ensure one, Mečiar had no choice but to stand aside and allow the elections to go ahead unhindered.

The results were emphatic, the more so on the back of an 84 per cent turnout and the failure of the Workers Party to cross the 5 per cent threshold. The combined opposition had a three-fifths majority. Although HZDS had just managed to squeeze in as the largest single party and some of its senior deputies said this meant they had therefore won the elections, Slovakia had made a clear and unambiguous vote for change. (See Table 8.1.)

Mečiar had lost power twice before but he had never been defeated at the polls. It took him a few days to recover his composure and on September 30, he came out into the open with an appearance on state television's *Ako d'alej pán premiér* (What next Mr Premier?). Some of his comments were revealing. He accused the people of Banská Bystrica and other areas of voting illogically in supporting the opposition when the government had been so generous to them in terms of investment. He also said his defeat marked the end of the 'Slovak Way', strongly implying that the new government would betray Slovakia abroad. Both of these reflected his style of government and the perception that had guided his dealings with opponents since the end of the federation.

But he did accept defeat and closed an extraordinary premiership by singing the lines of a traditional folk song. 'Farewell, I am leaving you. I did not hurt, I did not hurt a single one of you.'

There were those who disagreed with that last part, and others who questioned the state of mind of a leader who departed by singing a song. But this was unimportant. Mečiar had gone.

A glorious revolution

The four-party coalition government formed after the 1998 elections can be loosely compared with the broad conglomerate groupings which came together at the fall of communism. Its role as an unofficial joint opposition in the run up to polling day was formalised by a coalition agreement signed on October 28. But the energising force holding together this disparate collection of former communists, neoliberals, Christian democrats and ethnic Hungarians was clear: opposition to Mečiar and the style of government which he stood for. It is tempting to say that this is *all* that held them together. While this is broadly the right conclusion to draw, it should be borne in mind that this one unifying issue was *the* issue in Slovak political life.

As we move to a discussion of the difficulties faced by the current coalition it is important to set this against a broader picture of massive change from what went before. Let us recount the main charges against the Mečiar regime: The security services were abused for political purposes, including the kidnapping of the son of the country's head of state. The democratic system was compromised by the ejection of a dissident HZDS party member from parliament and threats to overturn the mandate of an opposition party, the Democratic Union. The principle of a law based society was disregarded by several high profile violations of the constitution. Civic peace was threatened by the government's generally hostile attitude to the Hungarian minority.[20] The state media were turned into a mouthpiece of the government. The privatisation process was used to enrich political allies and the ruling parties. The net result was Slovakia's exclusion from the European Union and NATO accession process, a dearth of foreign investment because of the justified perception of high political risk, and a deeply polarised and embittered domestic political scene.

It is only a small exaggeration to say that the election of a new government changed all that. If there is still deep hostility between the country's political parties, the government has at least given the opposition a presence on parliamentary committees. The European Union has congratulated the government on its moves to redemocratise society generally, including the drafting of a law on minority languages to placate the Hungarian minority. The elections came too late for Slovakia to rejoin the group of countries which became members of NATO in March 1999, but it has been publicly praised by the alliance for its cooperation during the Kosovo crisis. Relations with the Czech Republic and Hungary have improved dramatically. In order to try and ensure that the kind of aberrations of the Mečiar years are not repeated, criminal investigations have also been launched in connection with the activities of the security services in the Kováč kidnapping case and violations of the constitution arising out of the 1997 referendum. Ministries have drawn up so-called black books, detailing the abuses of the past. Some of the most obviously corrupt privatisations have also been partially reversed. This was all very well and it served to remind the coalition of why it had been created in the first place. But as one prominent political scientist[21] put it, black books looking at the past were one thing, what about white books dealing with the future?

New government, poisoned legacy

In the third quarter of 1998, economic growth in Slovakia was roaring ahead at an annual rate of 5.1 per cent. In the fourth quarter it plummeted to 0.5 per cent. The decline was partly explained by behavioural changes caused by the flotation of the Slovak crown on October 1. But as soon as the government began to examine the books it had inherited from the Mečiar government it became clear that there were massive underlying problems. Those economists who had warned that the impressively high headline growth rates since 1994 had been financed by heavy borrowing and reckless state spending on infrastructure were proved right in dramatic fashion. The fundamentals of the Slovak economy were dangerously weak.

This message was hammered home at the corporate level with the near collapse of the country's biggest company, VSŽ Holding a.s. Within weeks of the new government assuming office, VSŽ admitted that it had been unable to pay back a $35 million syndicated loan arranged by Merrill Lynch. Mečiar cronies were pushed aside and the true horror of the company's financial problems was revealed. In May 1999, VSŽ announced a net loss in the previous year of 11.1 billion crowns ($265 million), and this from a net profit the year before of 0.6 billion crowns.

The health of the country's second biggest company, oil refiner Slovnaft, was also brought into question by its own financial results.

From 1.8 billion crowns in 1997, net profits had fallen to just 70 million crowns. The fundamentals were better than at VSŽ but the scale of the decline was worrying none the less. More generally, with the depreciation of the Slovak crown, heavy company borrowing abroad was becoming increasingly expensive to service. An already shaky corporate base threatened to sink into oblivion and, perhaps, take the banking sector with it. Top economists were privately talking about the prospect of systemic collapse.

Presiding over this poisoned legacy, which included an unemployment rate of well over 15 per cent, was a government of four political groupings with widely varying ideological standpoints.

As unemployment persisted, the Party of the Democratic Left in particular would face growing demands from its constituency to change course from the austerity package announced in March 1999. The right–left divide represented the obvious potential for conflict. Less apparent is the internal logic of potential division within the right, which in the form of the SDK, had come together as a party just before the 1998 elections. This is a complex point but a brief discussion can usefully illuminate the full extent of the difficulties the current Slovak government faces.

The essential problem, which had shown itself in the vote on the government programme itself,[22] was not between the two main parties which had formed SDK – the Democratic Union and the Christian Democratic Movement–but between KDH and SDK itself.

We have seen that in the transition from communism, the key task facing all political parties was the creation of a voter base with solid party loyalty. Formed a matter of months after the 1989 revolution, the KDH has been more successful in this aim than any other party on the centre or the right of Slovak politics.

In this sense it had more to lose in joining SDK than the DU. Though the Prime Minister and head of SDK, Mikuláš Dzurinda, came out of KDH, Ján Čarnogursky remains the KDH leader. The logic is simple. Dzurinda has an interest in accruing voter loyalty to the SDK, which can be most easily achieved by the dissolution of voter loyalty to the KDH. The longer KDH remains a quiet force within the government the further voter loyalty can be expected to switch from it to the SDK of which it is a constituent part. Given that Čarnogursky has shown no intention of formally folding his party into the umbrella grouping with which it fought the elections, there is therefore a strong dynamic for division within the Slovak Democratic Coalition. This may explain why Čarnogursky has been prepared to take such a vocal and seemingly irresponsible stand on issues which appear to be secondary to the major problems confronting Slovak society.[23] The coalition within the coalition may yet prove to be the government's weakest point.

Dzurinda's coalition government exists for one purpose; to ensure that Vladimír Mečiar can never return to the helm of Slovak politics. With every year that passes Mečiar's chances recede. As the full extent of his past abuses becomes apparent his chances of attracting new voters to his cause become ever more difficult. At the other end of the demographic spectrum, his fiercely loyal but aged and dying voter base is being continually depleted.

Regardless of any problems within the SDK, the bonds holding the four-party coalition government together can be expected to slacken in proportion to the extent of Mečiar's decline. The May 1999 Presidential elections in which Mečiar took 42.82 per cent of the vote, losing to Rudolf Schuster with 57.18 per cent, suggest that the process may be a slow one. Mečiar was roundly beaten and support for his opponent was weakened by KDH concerns at Schuster's high ranking communist past.[24] But Mečiar showed yet again that he is a force to be reckoned with in Slovak politics.

Unless the government attempts to remove Mečiar by jailing him for abuse of power during his premiership, one likely scenario thus suggests itself. Mečiar's continued presence could continue to act as a unifying force for the broad coalition which unseated him in 1998. This need not mean the coalition parties remain in their current form. The SOP in particular appears to have no solid voter base to sustain itself and no set of policy initiatives to distinguish itself from the SDL, which can be expected to eventually swallow it up. But fear of a Mečiar comeback would be sufficient to hold the government together even in the face of its mounting economic problems.

In the short term – a period defined not by months or years but in terms of Mečiar's ability to seriously threaten a return to power – Slovak politics is likely therefore to remain deeply polarised but relatively stable as a result.

When Mečiar finally does depart from the political scene the HZDS party constructed around him is likely to disintegrate and its supporters to shift their loyalties elsewhere. This should not be a cause for concern. One major aim of the transition from communism was the establishment of a standard western style party-political system in which democratic parties left, right and centre vie for governmental power. That process was put on hold in Slovakia, and a second transition away from Mečiarism will be necessary before the transition from communism can be resumed. When, as it eventually must, the coalition collapses, there will be talk of crisis and division. It will in fact be a sign that Slovakia has grown up and shaken off the legacy of a man who did great harm to his people at home and disgraced his country abroad.[25]

Conclusion – Prospects for the EU

If communism represented the starting block from which the Czech and Slovak peoples broke free in 1989, membership of the European Union was the finishing line to which their leaders said they would take them. The consolation prize of NATO membership was bestowed on a not altogether grateful Czech Republic in early 1999 while Vladimír Mečiar saw to it that Slovakia was left on the sidelines. But the EU was the rich man's club everyone wanted to join. The prospect of membership represented the closest thing to an objective reference point by which progress in eliminating the baggage of four decades of communism could be judged. For the average citizen it meant elevation from the status of second class European, required to prove his honourable intentions at every border crossing to the West. For the businessman it offered up the hope of free access to some of the richest markets in the world. And for the politicians it loomed as both a carrot and a stick, promising a place among equals inside an emerging power block with real clout, and conversely the threat from the EU of public criticism if they failed to live up to the standards expected of them.

Asked to judge the preparedness of both countries for EU membership just after the middle of the last decade, most observers, including the author, would have put the Czech Republic somewhere near the top of the list and Slovakia inside the no hope zone at the bottom. By the end of the decade, the gap between the two countries appears to have narrowed drastically. Slovakia continues to rise from the ashes of Mečiarism while a growing realisation of the full extent of Václav Klaus's mismanagement of the reform process in his country has knocked the Czech Republic far from the pedestal so many were once prepared to put it on.

In so far as the transition from communism involved the move to a market economy and the creation of a democratic political environment it is possible to argue that the Czech Republic moves into the next decade with at least as many questions unanswered as its one-time federation partner. There is one key problem.

The Czech political system, unlike most others in the region, has not been tested under the strain of high unemployment and the social dislocation which arises from it. Czechs, Poles, Hungarians and Slovaks think they have suffered hardship since 1989. This is broadly true in all of the above countries apart from the Czech Republic, where unemployment stayed below 5.0 per cent for most of the decade and only in the late 1990s moved sharply up towards the 10 per cent level. Undoubtedly many pensioners have failed to sustain even the relatively low living standards they had under communism. Others have had to retrain, change jobs, and cope with the dislocating effects of inflation. But families have not often seen their main bread winner thrown out of a job with little hope of finding a new one. Their teenage sons have generally not left school with nothing to do but idle their days away in bitter rejection from a labour market which can't accommodate them. The failed Klausian economic reform project has postponed the day when real social pain would be felt. It has postponed it, moreover, to a time when calls for patience in the aftermath of a glorious revolution against communism could no longer be effectively used to mollify social discontent.

We have seen how the core failing of the Czech party political system has been its inability to produce viable coalitions with majority support in parliament. This was placed firmly within the context of a country in transition from communism. The left was unable to unite because of the presence of a hardline, unreconstructed communist party with which neither the Social Democrats nor the centrist KDÚ-ČSL could work. Despite sharing much in common ideologically, the right was unable to cooperate effectively precisely because of opposition within its ranks to Václav Klaus's political personality. It was argued that in the absence of strong loyalties to parties, and in the presence of a neo-liberal ideology which was difficult to sell to the masses, the temptation to construct the largest Czech party around the personality of Václav Klaus was too great to resist. The essential problem was that Klaus's particularly hard headed style proved unacceptable to the smaller rightist and centrist parties. The 'solution' to this utterly deadlocked alignment of forces was an extraordinary power sharing agreement struck between ODS and ČSSD and an attempt to change the electoral system to strengthen the hand of the larger parties. In so far as the European Union is concerned to see stability in the democratic process, this is bad news for the country's accession hopes. The electoral system faces radical change against a background of rising social tension. At the very time the EU is looking for safety, predictability and consistency, the Czech Republic is plunging into uncharted waters. Even if things go well, it is likely that at least one or two general elections would be needed to prove to the EU that the country's political system has matured to the point at which long term stability can be relied upon. It is also worth noting that any changes to the electoral system are likely to give greater power to the ODS and the ČSSD at precisely the time when their party leaderships appear to be turning ever more hostile to EU membership, adding in yet more uncertainty to an already unpredictable situation.

Victory over Vladimír Mečiar at parliamentary elections in 1998 and at presidential elections eight months later is unreservedly good news for Slovakia. But quite apart from Mečiar's vow to continue in politics, the Slovak political scene is also conspicuously immature. The battle lines since independence in 1993 have been drawn between supporters and opponents of Mečiar. In the wake of his defeat, the four-party government stretching from neo-liberal rightists to reformed communists on the left, is serving its country well, redemocratising society and showing courage in tackling deep seated economic problems. But the European Union can hardly consider inviting Slovakia to join its ranks as long as general elections are fought out between parties which are hostile to democracy and those who support it. When the coalition feels confident that Mečiar is no longer strong enough to mount a comeback, its own dissolution will mark the beginning and not the end of the road to becoming an advanced pluralistic democracy. Only after the democratic parties of left and right have fought out at least two more elections in the fashion to which western Europeans are accustomed can they become serious contenders for EU membership.

Looking at the party political scene alone, it would therefore seem highly unlikely that either the Czech or the Slovak Republics will be invited to join the European Union any time before 2005. The economy may pose equally difficult problems. In Chapters 5 and 6 we have seen how desperately the Czech Republic needs the full force of market reform to shake out the legacy of a communist era business culture and force companies to restructure. But it is difficult to see where the impetus for such courageous policies will come from. Social democracy will have to answer calls for a softer social approach from the growing ranks of the deprived. It is not likely to undertake a policy of rolling back the state on its own in any case. The right, dominated by Klaus, showed no serious inclination for translating free market rhetoric into reality in the 1990s and should not be expected to do so in the future. The Czech political scene appears to be drenched in a legacy of expectations about the role of the state which may even stretch beyond the country's communist past.

In Slovakia, the Dzurinda government has moved to address some of that country's most serious macro economic imbalances, but the banking and corporate sectors, hobbled by the systemic corruption of the Mečiar years, have barely begun to function like institutions of the same kind in the West.

Particularly in view of the mess most commentators made of the Czech Republic in the 1990s, the most sensible approach to predicting the prospects for the Czech and Slovak economies seems to be to remain prepared for the unexpected. On the optimistic side, the most obvious features of these economies is their large catch-up potential *vis-à-vis* the West. It may be that a critical mass of market reforms – a point at which new cultural attitudes to business dealings combined with a firmly established legal framework for private property and contract – will at some point be reached and allow for a sudden acceleration in growth. Massive gains in productivity would under this scenario translate into rapid (8–10 per cent) yearly growth, bringing the Czech and Slovak Republics quickly to the EU average. This may happen but it should not be expected. Czechs and Slovaks must break through several concentric circles of economic problems. In terms of its main export markets and the model to which the two countries aspire, Europe is a continent in relative economic decline. Commentators have often held up German economic success as the western European standard which they can and should be aiming for.

This is not as clear cut as it seems, however. Germany like the EU itself has deep economic and social problems, and it seems that the consensus driven approach which is credited with providing the country with so much prosperity may at the very least be inappropriate for the new world economy. At worst the big social problems associated with social democracy may have been disguised for many years by post-war construction. Once the very high rates of growth associated with this in the first 25 years after the war began to subside, the debilitating effects of interventionist economic policy gradually began to show themselves. Also Germany, which Czechs and Slovaks habitually do use as a measuring stick, is not the be all and end all of the west European economy. Britain, for example, suffered deep relative decline to other European countries for most of the post-war era until the reforms of the early 1980s. Portugal, Greece and Spain remain well behind the average standard of living in the European Union as a whole. Why should we expect the Czechs and Slovaks to move in the direction of German prosperity rather than emulate some of the less successful economies of the south of Europe? These economies show little sign of any (very rapid) catching up process with front runners inside the EU – though they can be expected to make up some ground; and they do not have the depressing legacy of decades of a fiercely anti-business climate fostered by communism.

The main point here is the very unpredictability referred to above. What if things go wrong, or at least fail significantly to improve? The European Union has enough problems to contend with without having to pay huge subsidies to help reconstruct countries which may not have the wherewithal to do the job themselves. The EU will simply not take the risk until it can be reasonably sure that the Czech and Slovak Republics are capable of establishing developed market economies with the real prospect of sustained growth and therefore stability.

Even if the arguments put forward in this book do not stand up and Klaus and Mečiar have, despite all the evidence to the contrary, presided over a successful economic and political transformation in the 1990s, there is one additional problem which may be even more difficult to cope with than the other two.

The national question was once described as the third prong in the 'triple transition' in so far as the break-up of the common state was seen as a complicating factor for the development of a pluralistic democracy and the move to a market economy. Seven years after the split, this issue has clearly receded into the background. But it has been replaced by another national question which almost ensures that neither the Czech nor Slovak Republic will join the European Union as full members with equal rights at any time in the foreseeable future.

The Roma, or Gypsy, minority in the Czech and Slovak Republics are among the most alienated ethnic groups in the industrialised world. Despite numbering in the hundreds of thousands in both countries, Roma are almost entirely absent from mainstream politics. Unemployment stands at over 70 per cent. In the major banks and companies in Bratislava and Prague it is truly exceptional to find a Roma employee, let alone one in a high ranking position. Roma rights groups report high levels of illiteracy and poverty. The threat of racist abuse, physical and verbal, is a reality so pressing that many Roma find themselves concentrated in villages and settlements away from white society.

While locals blame Roma for widespread criminality, there is little recognition that respect for the rules of the game is bound to be less forthcoming from a people who have not been invited to play on fair and equal terms. The Roma question is the number one civil rights issue in the Czech and Slovak Republics. But the sheer scale of the problem has profound implications for either country's chances of joining the European Union. We saw in Chapter 6 how the alienation described above prompted many Czech Roma to seek a better life in emigration. The problem is just as acute in Slovakia, where the British government was forced to re-impose visa requirements in 1998 after nearly 2,000 Roma applied for political asylum.

The point should be obvious. To the public, perhaps the most important aspect of full membership of the European Union is free movement of people across borders. The tacit assumption underlying this policy is that most people in western Europe are fully integrated within their own countries, and though some may move abroad for a number of years, most will remain in a home environment in which they feel comfortable.

The EU, racked by high unemployment, heavily strained social security systems and, frankly, still plagued by a certain degree of hostility to dark skinned ethnic minorities, is simply not going to open its doors to the Czech and Slovak Republics as long as the Roma problem persists. This is not to say that the EU should see things this way. It is simply to recognise that politicians will inevitably answer the call of their constituencies. Paranoia over the prospect of an influx of cheap labour from the east would probably be sufficient to prevent full and early EU membership in any event. The absolutely realistic supposition that the racially abused Roma minority would move west in significant numbers will encourage policy makers in the EU to look for reasons why the Czech and Slovak Republics should not be invited as full members. The most likely result is a kind of half-way house arrangement sometime during the second half of the decade in which they are invited to join without the right to free movement. At the popular level, this will be received with consternation (although the Czech and Slovak government are themselves likely to ask for exemptions on issues such as foreign property buying rights). Czech and Slovak politicians should view this as a challenge. The integration of the Roma minority is an issue of justice. But it will also come back to haunt mainstream society unless and until it can summon up the energy to find a way forward.

In conclusion, major questions still remain to be answered in matters of politics, economics and nationality. The Czech and Slovak Republics can either turn in on themselves and look for scapegoats or they can confront the problems head on. The prize of membership in the European Union is still there to be taken. It may, however, take a little longer to achieve than many in either east or west are yet willing to admit.

Notes

Introduction

1 See Carol Skalnik Leff, *The Czech and Slovak Republics – Nation Versus State* (Oxford, Westview Press, 1998).
2 The second wave of coupon privatisation was delayed slightly.

Chapter 1

1 This point has been well made by George Schopflin. One of the most important differences between the western countries and the absolutisms of the East including Russia was the extent of the separation of powers between church and state – its near absence in the East and its developing strength in the West. The western political consciousness split in tandem with this separation. When westerners looked up to those in power they saw at least the beginnings of distinct sources of authority. Subjects of the Czarist autocracy, which was fully integrated with the orthodox church, were held in thrall to just one. Division implied arbitration, doubt and eventually impartial judgement. Holism implied single-minded obedience. This perhaps goes some way to explaining the development of the rule of law and the rise of liberalism in the West and its conspicuous absence in the East.
2 It was precisely the inability of successive Czechoslovak governments of widely varying descriptions to achieve this aim that did in fact seal the country's fate seven decades later.
3 T.G. Masaryk, *Spisy*, vol. 2 (Prague: n.p., 1934), p. 78, cited in Leff, p. 26.
4 Early agreement by some Slovak parties to the unitary state may have been encouraged by the brief invasion of Slovakia by Béla Kun's Soviet regime in 1919. Revolutionary or not, the Hungarian masters were back. This was bound to have promoted a sense of panic. Conversely, when, with the help of Czech soldiers, Kun's government had been comprehensively defeated and the nature of the post-war settlement became clearer, autonomy minded demands could be more safely made.
5 The point was reinforced by the common practice abroad of adjectivalising the country's name simply to Czech. The eastern Slovak town of Košice was thus frequently referred to as the Czech town of Košice. This practice could only excite national sensitivity.
6 The party was legal until Munich. In contrast with many other countries in the region it could therefore agitate more or less freely. When Czechs and Slovaks encountered communist propaganda after the war, they were not hearing such ideas for the first time.

179

7 This of course was one of the main factors providing fertile ground for Slovak grievances against the Czechs in the lead up to the break-up of Czechoslovakia in 1993.

8 Czechs, unlike the Poles, did not have a strong history of anti-Russian feelings.

9 Nevertheless, Slovakia did achieve some limited concessions in the form of a National Council but its significance diminished as communist control tightened. There was also no corresponding Czech council, leading to a state of 'asymmetry' which suggested to many Slovaks that its significance was more token than real.

10 The issue is all the more sensitive, albeit at the subliminal level, now that the independent Czech Republic no longer has Slovakia to give the Slavic people added weight.

11 This section relies heavily on Bradley F. Abrams'seminal essay, 'Morality, Wisdom and Revision: The Czech Opposition of the 1970s and the Expulsion of the Sudeten Germans.' *East European Politics and Societies*, 9, no. 2 (Spring 1995).

12 Ibid. p. 237. The quotation inside is from Mlynárik's Theses on the Deportation of the Czechoslovak Germans in Odsun p. 71.

Chapter 2

1 Despite its name, the pact was not merely about non-aggression. The Soviet Union was accorded major territorial concessions including Eastern Poland and eventually the Baltic states of Latvia, Lithuania and Estonia.

2 The communist party was legal up until Munich. But its conversion to Russian style communist practices was by no means automatic. It drew on the radical traditions of Austria. This created tensions with the Comintern, which initially rejected its application for membership above all because of differences over Slovakia and ethnic German participation. It took until the end of the decade, a new and younger leadership and massive internal purges before a party of impeccably Leninist credentials had been established.

3 The communists launched a massive and successful recruitment drive in the aftermath of the Second World War. Party membership rose from around 40,000 to 2.7 million three years later.

4 See R. Conquest, *The Great Terror*, revised edition (Harmondsworth: Penguin Books, 1971).

5 With some exceptions, the trials and executions of Slánský and other leading communist being the most prominent, life in Czechoslovakia in the 1950s is more usefully compared in important respects to Brezhnev's Soviet Union. The atmosphere and the practice was deeply repressive but it was never genocidal.

6 It is of course this total form of control that gives rise to the expression totalitarianism.

7 Bureaucratisation was an endemic problem to all Soviet style regimes. From 1969 to 1989, and despite the purges in the months after the invasion, the number of administrative officials rose from just under 700,000 to over 800,000.

8 Public opinion also became a factor towards the end of the period prior to Dubček's ascent to power. Demonstrations in Strahov by students in late 1967 were ruthlessly suppressed by police. In a relatively compact environment like Prague, news of the suppression spread fast and there is anecdotal evidence that many people were angered at the authorities' behaviour.

9 A translation is contained at the end of Dubček's autobiography: A. Dubček, *Hope Dies Last* (London: HarperCollins, 1993).

10 Ibid. p. 148.

11 Ibid. p. 305.

12 One possible reason for the timing of the invasion was that the Extraordinary Thirteenth Congress of the Communist Party was set for late August. The fear may well have been that Dubček would move further on this matter and may have transformed the party beyond all recognition, spelling the death of any communist party recognisable as such to Moscow.

13 See George Leggett, *The Cheka: Lenin's Political Police* (Oxford: Clarendon Press, 1986).

14 'The whole Czechoslovak affair left the Soviet leaders with a profound distrust of intellectuals, especially in the humanities and social sciences, as well as of all economic reformers.' Geoffrey Hosking, *A History of the Soviet Union* (London: Fontana Paperbacks and William Collins, 1985), p. 374

15 The party purges also created the space for the party to tie new people into the regime through direct membership.

16 J. Patočka, *What Charter 77 Is and What It Is Not* (7 January 1977). In *Good-Bye Samizdat*, ed. Markéta Goetz-Stankiewicz (NorthWestern University Press, 1992), p. 143.

17 V. Havel, *The Trial* (October 1976). In *Open Letters, Selected Writings 1965–1990*, selected and edited by Paul Wilson (New York: Vintage Books, 1992), pp. 106–7.

18 See Misha Glenny, *The Rebirth of History* (London: Penguin, 1993).

19 G. Schopflin, 'The End of Communism in Eastern Europe.' *International Affairs*, 66 (1990): 3–16.

20 In the late 1980s this concern to retain some linkage to traditional ideology encouraged some extraordinary speeches from Gorbachev, suggesting that Lenin had in fact been a supporter of parliamentary democracy and freedom of speech.

21 Gorbachev later said that the Soviet leadership renounced the so-called Brezhnev doctrine which had been used to justify military intervention to sustain the Soviet empire in eastern Europe at the time of Communist Party general secretary Chernenko's funeral. He also claims that the inspiration for his reforms was indeed Dubček's Prague Spring.

22 Schopflin suggests (p. 231) that the central committee meeting of November 24 was told that the Interior Ministry could cope with demonstrations of up to 50,000 but not the 200,000–250,000 jamming Wenceslas Square at the time.
23 Dubček, p. 271.

Chapter 3

1 See IVVM December opinion poll, released to media on December 21. Fifty-five per cent of respondents said Havel should go.
2 Havlová caused a sensation when, following the president's re-election in 1998, Jan Vik of the Republicans made a highly critical speech of Havel and she put two fingers in her mouth and to the astonishment of assembled deputies let off a piercing cat whistle. Vik was quoted by ČTK as saying 'That whistle was so loud, the last time I'd heard a whistle like that was at the football . . . I turned around and thought "what the hell was that?"' Others thought it was the most interesting moment of the event and was a fitting response to the nonsense being blurted out by communist and republican figures. It wasn't exactly the normal reaction expected of the first lady, though, and a poll in February suggested that almost two-thirds of Czechs disapproved. Havel married her in Jan. 1997, around a year after his previous wife, Olga, died. Many thought this was indecent haste.
3 Some Solidarity activists in Poland said that it gave them a new sense of what they were doing and encouraged them in the face of their own oppressors.
4 V. Havel, 'The Power of the Powerless,' in *Open Letters, Selected Writings 1965–1990*, selected and edited by Paul Wilson (New York: Vintage, 1992), pp. 125–214.
5 Ibid. pp. 132–3.
6 Ibid. p. 136.
7 A supporting contrast also applies across the Soviet bloc and the varying degree of ideological orthodoxy of the regimes within it. Dissidence for instance was much more heavily concentrated among intellectuals in Czechoslovakia than in Poland, where the ideological possibilities of the regime were limited and compromised by the power and influence of the Roman Catholic church. The relatively healthier state of the Czechoslovak economy also made collaboration with the regime a more acceptable option in Czechoslovakia. Outbreaks of violence by the state against individuals and vice versa were considerably more common in Poland. And yet Poland was a much less repressive society in terms of the totalising nature of the rule the communist party tried to exercise. Czechoslovakia's rulers were able to suffocate dissident potential more successfully through the pervasiveness of ideology supported by the censorship and persecution of intellectuals which sustained it. Poland did this too but societal conditions made it less worthwhile. Actual violence and the threat of it

was a more pressing reality for society at large in Poland than in Czechoslovakia.

8 Ibid. p. 207.

9 New Year's Address to the Nation, Prague, 1 January 1990 in V. Havel, *The Art of the Impossible, Politics as Morality in Practice* (Fromm International, New York, 1998), pp. 3–9.

10 Also available on ČTK, 9 Dec. 1997.

11 Klaus, who was sitting in the middle of the front row, pointedly refused to applaud the end of the president's speech.

12 Full text quoted by ČTK 9 Dec. 1997.

13 Timothy Garton Ash, 'Prague: Intellectuals and Politicians,' *New York Review of Books* (12 January 1995), pp. 34–41.

14 We know that Havel signed a highly dubious lustration bill against his better judgement. We all hope that he signed the anti-Gypsy citizenship law in a similar spirit of unhappy resignation.

15 Havel's speech on 28 October 1998 on Czech National Day. Available on Prague Castle Internet site: www.hrad.cz

Chapter 4

1 The word 'crisis' here should be understood relative to the context of the country. At the highest levels of state power they appear to have had few problems. The country has a bicameral parliamentary system consisting of a powerful lower house, the Chamber of Deputies, where legislation is drafted and then passed on to a weaker upper house, the Senate. The Senate, which can draft laws of its own in some areas, can reject laws passed in the Chamber of Deputies but this can then be overridden by a simple majority in the lower house. Although there has been persistent opposition to the very existence of a second chamber, notably from Václav Klaus's Civic Democratic Party, the parliamentary system functions smoothly. To sum up the most basic points, free elections have been held among a plurality of parties with no objections to the results being raised from any serious quarter. Deputies have come and gone. The composition of governments has changed with little questioning of the legitimacy of the democratic process and no attempts (yet) to use constitutional or unconstitutional means to rig the system in favour of one party or another. There is a 5 per cent threshold required for parliamentary representation which is clearly designed to exclude very small parties from parliament, but size here is the only contingent factor and several developed democracies, notably Germany, employ similar safeguards. No powerful forces advocate anything other than the democratic set up which has already been established, and those that do tiptoe on the fringes of political life showing little sign of invading the mainstream.

2 The term has been used frequently by George Schopflin. The author is not aware whether he in fact coined it.

3 G. Schopflin, *Politics in Eastern Europe* (Oxford: Blackwell Publishers, 1993), pp. 259–60.

4 On the contrary, ODS has been consistently prepared to risk antagonising its centrist coalition partner the Christian Democratic Union–Czechoslovak People's Party (KDU-ČSL) over its ambivalence to the restitution of church property. Although, there were signs that ODS was starting to play on nationalist themes in the latter part of the decade, ethnic nationalism cannot be said to have formed an important aspect of mainstream right wing policy.

5 It also served to allow policy to translate into the creation of a new class which would eventually provide the right with the solid electoral constituency it lacked.

6 The most prominent critic of communism in western Europe, Margaret Thatcher, was herself a neo-liberal of sorts, giving Czech leaders a concrete example to follow.

7 This is a problem for all neo-liberal politicians in all countries. At the level of social and economic policy, left-leaning arguments are simpler and easier to sloganise. They are often inherently populistic.

8 Cyril Svoboda was quoted in *Právo* 24 March 1999 as saying he thought ordinary members of his party inclined more to the ČSSD than ODS.

9 The party appears to believe that the 1948 revolution was 'stolen' and that the wrong path was pursued. Nevertheless, despite professing support for parliamentary democracy it claims to agree with the main tenets of Marxism-Leninism – not an ideology known for its commitment to free elections or a pluralistic society.

10 This problem could also be said to have affected the communists themselves. The presence of the Social Democrats – a party with a long history – may have hindered reformist voices in the KSČM. If the communist had re-formed they would have been competing head on with the ČSSD.

11 See *Lidové Noviny*, 18 June (in English on ČTK the same day). An article quoting an unnamed former party member said senior Republican leaders regularly greet each other with the fascist salute and phrases such as 'sieg heil' or 'heil Hitler'. The source also said they openly discuss the genocide of the Gypsies.

12 Lux said in June 1997 that Klaus had actually hidden from the cabinet a report outlining how serious the economy's problems were Klaus said Lux was talking nonsense.

13 It is possible that this role has been exaggerated somewhat since neither Poland nor Hungary had really comparable figures and both managed to develop equally successful political democracies.

14 ODS Deputy Chairman Miroslav Macek. Quoted by ČTK 20 November 1997.

15 The author was present at the news conference where the comments were made.

16 'Following its party congress in Podebrady in December 1997, the ODS essentially became a sect: it became the party of Václav Klaus. Everything the party did between the Podebrady Congress and the early elections was

directly connected with Klaus's personally.' Jiří Pehe, in his essay 'Czech Crisis Deepens', *The New Presence*, January 1999.

17 Defeat for the pensioners occasioned one of the more gruesome episodes of the last 10 years. The party leader had promised to eat a bug if his party failed to get into parliament. He duly obliged.

18 Full text published by ČTK 9 July 1999.

19 Havel's statement on the agreement. See ČTK 9 July.

20. The extremely low turnout at the Senate elections later the same year was in fact heralded by many as evidence that alienation from politics had set in as a result. The strong showing of mainstream parties opposed to the agreement and the popularity of independent candidates at local elections was seen as supporting this point of view. The logic is clear but more evidence will be needed. The senate and the municipalities are far less visible than the lower house and voters consequently attach less importance to them. Still ČSSD at least was clearly having problems getting its supporters out to vote.

21 The ODS-ČSSD agreement called for constitutional proposals within 12 months, but at a time of rising unemployment and deep economic recession there are clear dangers in changing the electoral system. What happens if ČSSD or ODS support suddenly takes a downturn?

Chapter 5

1 V. Klaus, *Renaissance: The Rebirth of Liberty in the Heart of Europe* (The Cato Institute. Washington DC, 1997), pp. 3–5.

2 Industrial output in the first quarter of 1999 dropped 9.1 per cent year-on-year. Unemployment in April 1999 was 8.2 per cent. Source: Czech Statistical Office. See website: www.czso.cz

3 The European Commission's progress report on accession candidates, published 5 November, 1998, warned the Czech Republic that it had lost some ground. One specific concern was insufficient industrial restructuring.

4 Speech on 11 May 1996 at Žofín. It was not clear whether she said this in deference to her successor John Major or whether it was in fact a veiled reference to the esteem in which she held herself as premier in the 1980s.

5 Address delivered to the Mont Pélerin Society, Cannes, France, 26 Sept. 1994. *Renaissance*, p. 23.

6 His 10 commandments of economic reform suggest he may have set his sights somewhat higher, although to be fair, and in contrast, he did revisit them.

7 It is clear that Klaus was uncomfortable in a democratic environment. Analogies with Margaret Thatcher are way off the mark. Even at her most strident she never approached his level of what can only be described as contempt for criticism, however mild. The author was witness to many such occasions which bore this out.

8 Speech delivered at the fourth CEEPN annual conference on 'Privatisation in Central and Eastern Europe,' Ljubljana, Slovenia, December 1993. Quoted in Václav Klaus, *Renaissance*.

9 Chemapol and Junek provide the classic example.

10 'In 1990 it was recognised by the reformers that sweeping privatisation was necessary. Because of the limited savings of Czechoslovak citizens, standard privatization by sale to domestic citizens would have had to be protracted; 600 years was one estimate. Rapid sales at market prices would have had to mean sales to foreigners. This was also seen as undesirable.' Vladimír Dlouhý and Jan Mládek, 'Privatization and Corporate Control in the Czech Republic'. In *Economic Policy* (Dec. 1994), pp. 156–7.

11 *Renaissance*, pp. 73–4.

12 Restructuring prior to privatisation was dismissed as an extension of the propensity of social democratic governments to 'pick winners'.

13 *Renaissance*, p. 71.

14 It might if you can wait 500 years, but even then laws will almost certainly have to back up whatever institutional arrangements emerge in order to deal with conflicts of interest.

15 Marxist and communitarian scholars have long recognised this. The neo-liberal state may be minimal in scope but it must be powerful in securing the property rights which underpin it. Many of the less sophisticated left-leaning criticisms of neo-liberalism also miss the point spectacularly, and it could be that Klaus, whose early political upbringing was amongst many such people, fell for the bait they threw him. Free market capitalism, despite many such assertions in this direction, is anything but the law of the jungle. It depends on highly complex legal forms and a culture to underpin them.

16 *Renaissance*, p. 26. This is about as close as it is possible to find Klaus getting to a recognition of the importance of property rights but, typically, the point once raised is immediately dropped. Surveying the collection of speeches and essays in *Renaissance* and other publications, such arguments where they do appear are often nothing more than one liners in the middle of other arguments perceived to be of greater importance.

17 Quoted by ČTK 12 Dec.

18 Ibid.

19 *Prague Tribune*, 6 Dec. 1997.

20 ČTK 29 Aug. 1997. Speech to the Annual Economic Symposium on Policy Issues in Jackson Hole, Wyoming.

21 M. Bohatá, 'Some Implications of Voucher Privatisation for Corporate Governance', *Prague Economic Papers*, 1 (University of Economics, Prague, March 1998).

22 It is of course true that much the same sort of situation arises in the Austro-German banking system which has successfully overseen two of the world's richest and most successful economies. The key problem, it could therefore be argued, was more to do with whether the banks could assess credit risk properly. The Anglo-American model, separating owners from lenders, would not necessarily have resulted in a more efficient

allocation of funds unless the bank employees knew what they were doing. The obvious way around this would have been to privatise the banks to foreign owners as a first step in the reform process. The government was probably concerned that this would lead to a massive wave of bankruptcies. But this could have been met by simultaneously allowing foreigners to take large stakes in the country's biggest companies. Foreign banks would then have been more confident in the ability of corporates to restructure their businesses.

23 They were also of course storing up a mountain of unemployment which did in fact begin to show itself in the last two years of the decade. The political price would be paid by the Zeman government which governed with Klaus's support. By this stage, as will be argued, political considerations were paramount for Klaus.

24 Josef Poeschl, *Czech Republic: 'An Economy Out of Steam'* (Feb. 1999), p. 1. Published by The Vienna Institute for Comparative Economic Studies. Poeschl also argues that internal commercial debt partially passed on from the communist past was a major problem. Companies ended up being so highly geared that they had to spend a vast proportion of any profits they did make on servicing debt. This made innovation and further investment less likely. He suggests that while companies and banks were in state hands it may have been sensible to write off this debt, which would have been purely an accounting matter at the time.

25 See M. Dangerfield, 'Ideology and the Czech Transformation: Neoliberal Rhetoric or Neoliberal Reality?', *East European Politics and Societies* (Fall 1997).

26 I am using neo-liberal, New Right and neo-conservative more or less interchangeably. Strictly speaking, the New Right comprises both the neo-liberals and the neo-conservatives, the latter two groups sometimes holding radically different views. Nevertheless, Klaus himself is anything but consistent in the use of these terms, sometimes referring to himself as neo-liberal and other times as a sort of conservative. The point at issue here is whether Klaus can be placed anywhere on the New Right.

27 V. Klaus, 'The Ten Commandments of System Reform', quoted in *Prospects for the Czech Republic* (APP Group a.s., David and Shoel s.r.o., COWI in cooperation with other firms, 1994, Prague).

28 Nevertheless, it is possible to argue that the devaluation was actually far too heavy. Inflation in Czechoslovakia had more to do with a cost push, associated with the need to meet international prices for commodities, rather than demand pull as the government appeared to assume. Czech companies were therefore hit from both sides by heavy devaluation raising still further the costs of many inputs, and simultaneously a tight monetary policy squeezing demand in their own economy. Klaus may have misinterpreted the underlying reason for inflation in his economy. Also, western machinery was made very expensive. Social democratic economists have argued that it would have been better to keep devaluation lower and keep some duties.

29 This was partly fortuitous, since the country had a much smaller inflationary overhang and its industry was arguably more efficient. From Klaus's longer term point of view, it was doubly fortuitous that most of the really unpleasant consequences were felt in Slovakia and not in the Czech lands at all. Unemployment had already risen to close to 12 per cent in Slovakia compared with a Czech jobless rate of little more than 4 per cent.

30 The evidence showed that if any country in the region could quickly soak up the body blows of rapid macro economic stabilisation it was Czechoslovakia. Klaus knew this and there is little doubt that he represented the forces of progress at this time. The alternatives were, generally speaking, just as Klaus described them; unadventurous, complacently slow and insufficiently fired up with the energy necessary to push through the task at hand.

31 M. Ornstein,'Václav Klaus: Revolutionary and Parliamentarian', *East European Constitutional Review* (Winter 1998), p. 47.

32 H. Appel, 'Justice and the Reformulation of Property Rights in the Czech Republic', *East European Politics and Societies*, 9 (Winter 1995), p. 36.

33 The EBRD's 1995 Transition report, pp. 181–4, highlights in more detailed form the dangers of taking the GDP figures at face value.

34 Figures available in EBRD transition reports.

35 Conversely of course, the companies were unable to compete on foreign markets despite the advantage of much lower costs of production.

36 The author was working as a financial journalist in Prague at the time. These impressions are based on his recollection of dozens of conversations with foreign and domestic investors and share dealers. While analysts throughout the mid to late 1990s were writing reports to investors and telling financial journalists of big potential gains in the major industrial companies, they were simultaneously able to expound on problems of corporate governance, the incestuous relationship between the companies and the banks and the general lack of corporate restructuring. But they appeared unwilling to factor these concerns into their predictions on the stock market or the economy as a whole. Some now admit that it was simply not in their interests to do so. Their jobs and salaries depended on good news or prospects thereof going to their bosses in London and New York.

37 Imagine how much different Margaret Thatcher's record on unemployment would have looked if the north of England had seceded in the early 1980s.

38 Standard and Poor's, press release, 5 Nov. 1998.

Chapter 6

1 STEM poll. Published by ČTK 20 October 1998.

2 Sofres-Factum poll. Quoted by ČTK 5 May 1998.

3 Lizner was the head of the finance ministry's centre for voucher privatisation which conducted the auction of shares in return for vouchers

contained in the voucher books made available to all adults for a small sum. In Chapter 5 it was noted that the bidding system turned vouchers into surrogate money. This organisation processed the orders in the bidding process. But Lizner was also the head of the Central Securities Registry which kept records of all share transactions and the accounts of the shareholders.

4   Q. Reed, 'Political Corruption, Privatisation and Control in the Czech Republic: a Case Study of Problems in Multiple Transition.' Doctoral thesis (Oriel College, Oxford, Sept. 1996), p. 275.

5   He was released in December 1998 having served three years.

6   *Mladá Fronta Dnes*, 5 Feb. 1996. Available in English from ČTK on same day.

7   As examples: OKD was fully state owned and IPB Bank was 45 per cent state owned.

8   The audit was released by ODS on 13 May, 1998. It said that financial disclosure rules had been ignored and that accounting records had been 'systematically' distorted.

9   There could be cause and effect here. ODA's voters are the most educated of all the parties' and may have felt disgusted by the affair. However, it is difficult to draw such a conclusion since ODS split and US was formed too close to this event to draw firm conclusions about the way its supporters were thinking.

10 A crucial element in this equation was the near total state ownership of property. A one-party state which owned all the buildings had an automatic enforcement mechanism preventing anyone from starting up on their own.

11 It should be recognised that learning the necessary modes of behaviour as members of voluntary independent organisations is far more efficient than at elections themselves. Imagine how long it would take if they only had the once-in-every-four-years experience of a general election.

12 The Law on Resistance Against and the Illegitimacy of the Communist Regime was signed by Havel on 22 July 1993 having been passed in parliament by a big majority two weeks before.

13 V. Cepl, 'Ritual Sacrifices', *East European Constitutional Review*, 1 (Spring 1992), pp. 24–6.

14 J. Šiklová, 'Lustration, the Czech Way of Screening', *East European Constitutional Review* (Winter 1996), pp. 57–62.

15 Poll conducted by IVVM in February 1999.

16 Štepan has shown no signs of remorse, attempting to organise a hardline communist party since his release. In an interview with the author prior to the 1996 elections he still maintained that the whole revolution had been organised by the CIA, Mikhail Gorbachev and Germany and was supported only by a small minority of the people.

17 This part of the chapter is built around the personal recollections of the author himself who was working as a journalist in Prague for a large part of the 1990s.

18 Since *Telegraf* was a deeply loss making venture, Reed (pp. 283–4) has also raised the question as to whether the most pro-government newspaper may not have been being subsidised by taxpayers' money.

19 Scott Macmillan, *Prague Business Journal*, issue 12, 1997.

20 A demonstration outside the Economics University in Prague was estimated at several thousand strong. It was also broadcast live on television, another indicator of how seriously the tragedy was being taken.

21 IVVM, December 1996. Quoted by CTK, 11 December 1996.

22 Mayor Ladislav Hruška was quoted by Reuters, 2 June 1998.

23 See Jiřina Šiklová and Marta Milusakova, 'Denying Citizenship to the Czech Roma', *East European Constitutional Revue*, 7, no. 2 (Spring 1998).

24 Ibid. p. 62.

25 Letter to Václav Klaus, 22 April 1997, from the Commission on Security and Cooperation in Europe.

Chapter 7

1 M. Glenny, *The Rebirth of History. Eastern Europe in the Age of Democracy*, 2nd edn (Harmondsworth: Penguin, 1993), p. 246.

2 S.J. Kirschbaum, *A History of Slovakia, The Struggle for Survival* (London: Macmillan, 1995), p. 211.

3 Ibid. pp. 200–3.

4 'The Response of Slovak Historians to M.S. Ďurica's Book: A History of Slovakia and the Slovaks', in *Kritika Kontext* (2–3/97), pp. 34–40.

5 The view of the Movement for a Democratic Slovakia (HZDS); ibid. p. 63.

6 Kirschbaum, p. 196.

7 Ibid. p. 187.

8 Ibid. p. 230.

9 Most easily found on internet page www.government.gov.sk

10 From personal recollection, there were many conversations in Prague in 1992 and 1993 when it was almost impossible to find a Czech who thought the split was a good idea. Most had no comprehension of why Slovaks wanted independence or even greater autonomy, and in my opinion simply thought they were being bloody minded. Western journalists tended to share the same opinion.

11 P. Pithart, 'Towards a Shared Freedom 1968–89', in *The End of Czechoslovakia*, ed. Jiři Musil (Budapest: Central European University Press, 1995), pp. 221–2.

12 Some of the more hysterically anti-Czech nationalists have suggested there was no difference between Hungarian domination and acceptance of the Czech agenda, which is plainly absurd.

13 J. Rychlík, 'National Consciousness and the Common State (A Historical-Ethnological Analysis)', in Musil ed.

14 Novotny's insensitive handling of the Slovak question contributed to his own downfall. He managed to alienate potentially important conservative figures in the Slovak Communist Party leadership.

15 A. Innes, 'The Break up of Czechoslovakia: The Impact of Party Development on the Separation of the State', *East European Politics and Societies*, 11,
no. 3 (Fall 1997), pp. 393–435.
16 Innes, p. 433, says: 'In both Republics, programs originated from the
top down, bypassing the majority preference for a common state
altogether, complex as these latter preferences were.' Describing the
preferences as complex as a codicil simply won't wash. The complexity
was the issue.
17 Ibid. p. 431.
18 There could only be one federal prime minister, one federal foreign minister and so on down to the most hopeful party functionary eyeing a
future political career.
19 Innes, p. 433.
20 This is not to say that free market reform was not in the interests of
Slovakia but was in the interests of the Czech Republic. It is simply
to acknowledge that different economic bases made some policy
initiatives easier to implement in terms of popular support in either
country.
21 Innes, p. 394.
22 Allison K. Stanger, 'Czechoslovakia's Dissolution as an Unintended
Consequence of the Velvet Constitutional Revolution', *East European Constitutional Review* (Fall 1996), pp. 40–6.
23 Ibid. p. 42.
24 Ibid. p. 42.
25 In so far as we are concerned with the transitional nature of the power of
Havel's office – that is, a moral authority derived specifically from his role
under and flowing out of communism – it is also important to recognise
that while it may have been overwhelming in the Czech lands it was far
from certain in Slovakia.
26 A largely ceremonial presidency run by Havel, it can be argued, was useful
and appropriate for the Czech Republic, but the Masarykian traditions
with which it was associated may actually have contributed to a Slovak
sense of not really being part of the same body politic. In endowing the
newly democratised institutions of the Czechoslovak federation with the
humanistic, and largely secular, values which Havel represented he could
not help but endow them with something quintessentially Czech. For
Czechs this was of great value for precisely the same reasons that it may
actually have contributed to a Slovak sense of alienation. It is just possible
therefore that in the specific conditions of transition from communism
Havel was particularly inappropriate as a president who wanted to preserve the federation.
27 Slovak intellectuals and the Hungarian minority were the most concerned. The Hungarians had seen Czech influence as a restraining factor
on Slovak hostility to them. Intellectuals, who had little feeling for the
national question in any case, were unhappy at losing Prague, a cultural
and intellectual centre of real stature in Europe.

Chapter 8

1 *Slovenská Republika* on 4 February 1998 described the comment as 'grossly offensive and insolent'.

2 Alexander Dubček, then chairman of the Czechoslovak federal assembly, described the sacking as a mistake. Meanwhile, Bratislava radio carried an interview with the Slovak TUC chairman Svetozár Korbel, who said that a five-minute strike in protest at the Slovak Parliament Presidium's changes in the government would be staged on 26 April. BBC Monitoring Service, April 25.

3 Mečiar had always been deposed before the end of any term of office, always contesting elections without having to defend a record in office. In the difficult transformation period, he was handed the initiative in being able to make promises while simultaneously attacking the government from the opposition.

4 Party leader Ján Slota once famously suggested that the solution to the Roma question was a 'small yard and a long whip.' In a drunken outburst in 1999 he even called on his compatriots to 'get in the tanks and go and flatten Budapest.'

5 The SDL was the product of those intellectual forces in the middle and lower levels of the Slovak Communist Party which wanted to develop left wing ideas along the lines of mainstream European social democracy. It was formed by a professional historian, Peter Weiss, who was fully representative of the youngish intellectuals and professionals who joined him. Its mass voter base was clearly located in the industrial working class and those who had been driven into unemployment by loss making factory closures. In answering the call of this constituency, the SDL offered conditional support for privatisation, seeking to keep natural monopolies and strategic companies in state hands and ensure a strong social safety net. This agenda set the SDL apart from the other parties in the opposition, allowing Mečiar to play their concerns about privatisation off against the other parties hostile to him until he overplayed his hand in this regard in the middle of 1996. The SDL joined the government of Jozef Moravčík in the period between Mečiar's fall in March 1994 and the elections which that government lost later the same year.

6 The ZRS party leader Ján Lupták voiced the concerns of a sizeable minority within the SDL which believed the party to be dominated by intellectuals and not sufficiently in touch with its working class base. Since SDL had joined the Moravčik government in 1994, the ZRS could also set itself apart from the SDL which had presided over the high unemployment racking Slovakia since the beginning of the decade. This created the space for splitting the left wing vote and electoral success in the 1994 elections.

7 The Democratic Union (DU) was formed mainly by defectors from Mečiar's HZDS and was characterised by opposition to the corruption and populism of the party they left. It was initially led by Moravčík, who

had been Mečiar's foreign minister, and in 1996 joined the Liberal International. It saw itself as the successor to the VPN.

8 Ján Čarnogurský described this as putting Slovakia on the level of a banana republic.

9 Gaulieder was in a position to know what the SIS was up to since he chaired a parliamentary committee overseeing its operations.

10 Mečiar took exception to this suggestion and lodged an official complaint against the Austrian authorities.

11 Britain's *Independent on Sunday* asked its readers to imagine that Prince Charles had been kidnapped at gun point, forced to drink a bottle of liquor and dumped outside a French police station. The chief suspect, Prime Minister John Major.

12 The Christian Democratic Movement (KDH) was founded in Nitra in February 1990. Its leader, Ján Čarnogurský, a prominent Slovak dissident who was jailed under communism, built KDH into an anti-communist, catholic party with distinctly conservative leanings. In contrast with the neo-liberal parties in the Czech Republic, the KDH exemplified more clearly the problems faced by right wing parties emerging from communism. In a devoutly Catholic country KDH had obvious appeal to some sections of the population. But overplaying this theme appears to have alienated many, especially the young, who had no natural affinity with what appeared to be an old-fashioned style of politics. Nevertheless, Čarnogurský's moral authority was sufficient to sustain the party as a leading alternative to HZDS. KDH had been in coalition with Mečiar since the 1990 elections and Čarnogurský replaced him as premier after his first ouster in 1991 until elections in 1992.

13 EU Commission summary on Slovakia's fitness to begin European Union membership talks. Issued as press release from Brussels, 16 July 1997.

14 In June 1996, the SDL had offered help to Mečiar during a coalition crisis in return for a promise to halt sell-offs of the country's banks. But once Mečiar had re-established links with his partners he reneged on the agreement, thus ensuring that the one potential weak link in the forces ranged against him would never trust him again.

15 The other major force in the parliamentary opposition came from the Hungarian minority. What eventually became the Hungarian Coalition was formed to represent the interests of the country's 500,000 ethnic Hungarians. In view of the Mečiar government's generally hostile stance towards the Hungarians it was a natural partner for the other mainstream parties in opposition. However, fear of being labelled disloyal to the state discouraged close cooperation until relatively late on.

16 See Reed, pp. 275–7.

17 It also forbade party election broadcasts during the campaign on anything other than the state media. This was designed to force voters to follow the election on a state radio and television service which was firmly under Mečiar's control.

18 In the end this extraordinary list of celebrities had little more effect than to spark off a nation-wide spate of blond jokes and convince the electorate that a large pot of cash could convince some people to appear with anybody.

19 Interviewed by the author in Bratislava, May 1999.

20 In September 1997 Mečiar had even raised the prospect of population transfers in which any ethnic Hungarians wanting to do so should be accepted 'back' in the Hungarian motherland.

21 Miroslav Kusý, of the Comenius University in Bratislava. Interviewed by the author in early 1999.

22 Čarnogurský, Justice Minister in the new government, voted against the government programme because he believed it had failed to give sufficient weight to a prior agreement on financing of religious schools.

23 The most infamous was his decision to vote against the programme of the government of which he is a minister for the reasons cited in note 22..

24 Schuster was a member of the Central Committee of the Slovak Communist Party in 1989.

25 Slovakia's poor image abroad is the subject of an excellent discusion by Beblavý and Salner: *Tvorcovia Obrazu Obraz Tvorcov* (Bratislava: Centrum Pre Spoločenskú A Meciálnu Analýzu, 1999).

Bibliography

B.F. Abrams, 'Morality, Wisdom and Revision: The Czech Opposition of the 1970s and the Expulsion of the Sudeten Germans.' *East European Politics and Societies*, vol. 9, no. 2 (Spring 1995).

H. Appel, 'Justice and the Reformulation of Property Rights in the Czech Republic', *East European Politics and Societies*, vol. 9, no. 1 (1995) pp. 22–40.

T.G. Ash, 'Prague: Intellectuals and Politicians,' *New York Review of Books* (12 January 1995), pp. 34–41.

T.A. Baylis, 'Elite Change After Communism: Eastern Germany, the Czech Republic and Slovakia', *East European Politics and Societies*, 12, no. 2 (Spring 1998).

M. Beblavý and A.Salner. *Tvorcovia Obrazu Obraz Tvorcov* (Centrum Pre Spoločenskú A Meciálnu Analýzu, Bratislava, 1999).

M. Bohatá, 'Some Implications of Voucher Privatisation for Corporate Governance' *Prague Economic Papers*, vol. 1 (University of Economics, Prague, March 1998), pp. 59–65.

P. Bren, 'Lustration in the Czech and Slovak Republics', *RFE/RL Research Report*, 2, no. 29 (16 July 1993), pp. 16–22.

V. Cepl, 'Ritual Sacrifices', *East European Constitutional Review*, 1 (Spring 1992), pp. 24–6.

R. Conquest, *The Great Terror*, rev. edn (Harmondsworth: Penguin Books, 1971).

D. Conway, *A Farewell To Marx* (Harmondsworth: Penguin, 1987).

R. Cox and E. Frankland, 'The Federal State and the Breakup of Czechoslovakia: An Institutional Analysis', *Publius: The Journal of Federalism*, 25, no. 1 (1995), pp. 71–88.

M. Dangerfield, 'Ideology and the Czech Transformation: Neoliberal Rhetoric or Neoliberal Reality?', *East European Politics and Societies* (Fall 1997), pp. 436–69.

V. Dlouhý and J. Mládek. 'Privatization and Corporate Control in the Czech Republic', *Economic Policy* (Dec. 1994), pp. 156–70.

A. Dubček, *Hope Dies Last* (London: HarperCollins, 1993).

J. Elster, 'Consenting Adults or the Sorcerer's Apprentice?', *East European Constitutional Review*, 4, no. 1 (1995), pp. 36–41.

E.J. Evens, *Thatcher and Thatcherism* (London: Routledge, 1997).

M. Friedman and F. Friedman, *Free to Choose* (New York: Avon Books, 1979).

M. Glenny, *The Rebirth of History. Eastern Europe in the Age of Democracy*, 2nd edn (Harmondsworth: Penguin, 1993).

J. Gray and D. Willets, *Is Conservatism Dead?* (London: Profile Books and the Social Market Foundation, 1994, 1997).

D.G. Green, *The New Right* (Brighton: Wheatsheaf Books, 1987).

A. Grzymala-Busse, 'Reform Efforts in the Czech and Slovak Communist Parties and Their Successors, 1988–1993', *East European Politics and Societies*, 12, no. 3 (Fall 1998), pp. 442–71.

V. Havel, *The Art of the Impossible, Politics as Morality in Practice* (New York: Fromm International, 1998).

V. Havel, *Disturbing the Peace* (New York: Alfred A. Knopf, 1990).

V. Havel, *Letters to Olga* (London: Faber and Faber, 1998).

V. Havel, *Open Letters, Selected writings 1965–1990*, selected and ed. Paul Wilson (New York: Vintage Books, 1992).

G. Hosking, *A History of the Soviet Union* (London: Fontana Paperbacks and William Collins, 1985).

M. Ignatieff, *Blood and Belonging* (London: Vintage, 1994).

A. Innes, 'The Break up of Czechoslovakia: The Impact of Party Development on the Separation of the State', *East European Politics and Societies*, 11, no. 3 (Fall 1997), pp. 393–435.

P. Johnson, *Modern Times* (New York: HarperPerennial, 1992).

P. Kennedy, *The Rise and Fall of the Great Powers* (London: Fontana Press, 1988).

S.J. Kirschbaum, *A History of Slovakia, The Struggle for Survival* (London: Macmillan, 1995).

V. Klaus, 'The Ten Commandments of System Reform', quoted in *Prospects for the Czech Republic* (Prague: APP Group a.s., David and Shoel s.r.o., COWI in cooperation with other firms, 1994).

V. Klaus, *Renaissance: The Rebirth of Liberty in the Heart of Europe* (Washington, DC: The Cato Institute, 1997).

V. Klaus and T. Ježek, 'Social Criticism, False Liberalsim, and Recent Changes in Czechoslovakia', *East European Politics*, 5, no. 1 (Winter, 1991), pp. 24–40.

E.M. Leeds, 'Voucher Privatization in Czechoslovakia', *Comparative Economic Studies*, 15, no. 3 (Fall 1993), pp. 19–37.

C. Skalnik Leff, *The Czech and Slovak Republics – Nation Versus State* (Oxford: Westview Press, 1998).

C. Skalnik Leff, 'Dysfunctional Democratisation? Institutional Conflict in Post-Communist Slovakia,' *Problems of Post-Communism*, 43, no. 5 (1996), pp. 36–50.

George Leggett, *The Cheka. Lenin's Political Police* (Oxford: Clarendon Press, 1986).

J. Musil, ed., *The End of Czechoslovakia* (Budapest: Central European University Press, 1995).

A. Nove, *An Economic History of the USSR* (Harmondsworth: Penguin, 1986).

M. Ornstein, 'Václav Klaus: Revolutionary and Parliamentarian', *East European Constitutional Review* (Winter 1998), pp. 46–55.

J. Pehe, 'Czech Crisis Deepens', *The New Presence* (Jan. 1999), pp. 10–11.

P. Pithart, 'Towards a Shared Freedom 1968–89', in *The End of Czechoslovakia*, ed. Jiři Musil (Budapest: Central European University Press, 1995).

J. Poeschl, 'An Economy Out of Steam' *Czech Public*, Feb. 1999, p. 1. Published by the Vienna Institute for Comparative Economic Studies.

R.B. Pynsent, *Questions of Identity: Czech and Slovak Ideas of Nationality and Personality* (London: Central European University Press, 1994).

Q. Reed, 'Political Corruption, Privatisation and Control in the Czech Republic: a Case Study of Problems in Multiple Transition.' Doctoral thesis, Oriel College, Oxford, 1996.

A. Ryan, *Property* (Milton Keynes: Open University Press, 1987).

J.Rychlík, 'National Consciousness and the Common State (A Historical-Ethnological Analysis)', in Musil, ed.

G. Schopflin, 'The End of Communism in Eastern Europe', *International Affairs*, 66 (1990), pp. 3–16.

G. Schopflin, 'Post-Communism: Constructing New Democracies in Central Europe', *International Affairs*, 67, no. 2 (1991), pp. 235–50.

G. Schopflin, *Politics in Eastern Europe* (Oxford: Blackwell Publishers, 1993).

Jiřina Šiklová and Marta Milušaková, 'Denying Citizenship to the Czech Roma', *East European Constitutional Review*, 7, no. 2 (Spring 1998).

J. Šiklová, 'Lustration, the Czech Way of Screening', *East European Constitutional Review* (Winter 1996), pp. 57–62.

Members of the Scientific Board of the History Section of the Slovak Academy of Science, 'The Response of Slovak Historians to M.S. Ďurica's Book: A History of Slovakia and the Slovaks', in *Kritika Kontext* (2–3/97), pp. 34–40.

M. Sommer, *Living in Freedom* (San Francisco: Mercury House, 1994).

A. K. Stanger, 'Czechoslovakia's Dissolution as an Unintended Consequence of the Velvet Constitutional Revolution', *East European Constitutional Review* (Fall 1996), pp. 40–6.

O. Ulč, 'Czechoslovakia's Velvet Divorce', *East European Quarterly*, 30, no. 3 (Sept. 1996), pp. 331–52.

G. Wightman, 'The Development of the Party System and the Break-up of Czechoslovakia', in Wightman, ed., *Party-Formation in East-Central Europe: Post-Communist Politics in Czechoslovakia, Hungary, Poland and Bulgaria.* (Brookfield, Vt.: Edward Elgar, 1995).

S. L. Wolchik, 'The Politics of Ethnicity in Post-communist Czechoslovakia', *East European Politics and Societies* (Winter 1994), pp. 153–88.

S.L. Wolchik 'The Czech Republic: Havel and the Evolution of the Presidency since 1989', in *Post Communist Presidents*, ed. Ray Taras (Cambridge: Cambridge University Press, 1997).

S.L. Wolchik, *Czechoslovakia in Transition: Politics Economics and Society* (London: Pinter Publishers, 1991).

Index

Printed in the United States
By Bookmasters